PERGAMON INTERNATIONAL POPULAR SCIENCE SERIES

THE MICROSCOPE

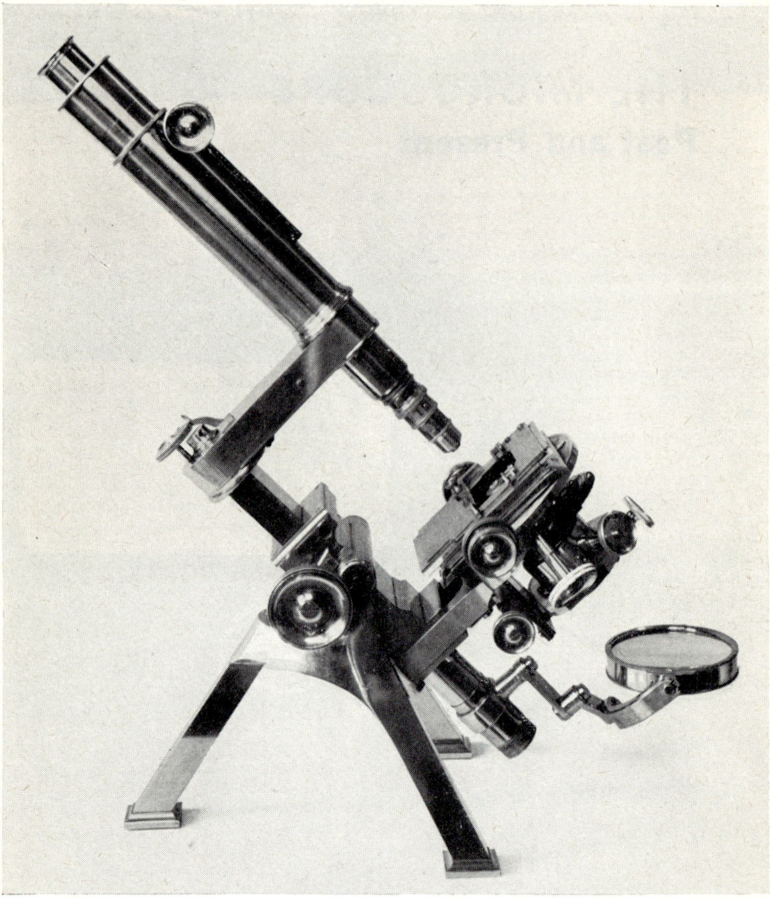

The Powell and Lealand "No. 1" microscope. This instrument represented perhaps the height of microscope design and craftsmanship in this country during the last century.

THE MICROSCOPE
Past and Present

BY

S. BRADBURY, M.A., D.Phil., F.R.M.S.
Fellow of Pembroke College, Lecturer in Human Anatomy, University of Oxford.

1966
THE QUEEN'S AWARD TO INDUSTRY 1966

PERGAMON PRESS

OXFORD · LONDON · EDINBURGH · NEW YORK
TORONTO · SYDNEY · PARIS · BRAUNSCHWEIG

PERGAMON PRESS LTD.,
Headington Hill Hall, Oxford
4 & 5 Fitzroy Square, London W.1
PERGAMON PRESS (SCOTLAND) LTD.,
2 & 3 Teviot Place, Edinburgh 1
PERGAMON PRESS INC.,
44–01 21st Street, Long Island City, New York 11101
PERGAMON OF CANADA LTD.,
207 Queen's Quay West, Toronto 1
PERGAMON PRESS (AUST.) PTY. LTD.,
19a Boundary Street, Rushcutters Bay, N.S.W. 2011
PERGAMON PRESS S.A.R.L.,
24 rue Écoles, Paris 5e
VIEWEG & SOHN GmbH,
Burgplatz 1, Braunschweig

First Edition 1968 PERGAMON INTERNATIONAL POPULAR SCIENCE SERIES
Library of Congress Catalog Card No. 68–24061

Printed in Great Britain by A. Wheaton & Co., Exeter

08 012848 3

For Sheila

In commendation of ye Microscope

Of all the Inventions none there is Surpasses
the Noble Florentine's Dioptrick Glasses
For what a better, fitter guift Could bee
in this World's Aged Luciosity.
To help our Blindnesse so as to devize
a paire of new & Artificial eyes
By whose augmenting power wee now see more
than all the world Has ever doun Before.

HENRY POWERS 1664

Contents

Preface

THE microscope is an instrument which is used at some time or another by almost every scientist; it is surprising, therefore, how little is known either about the best methods of using this valuable tool or about how it has developed from the earliest days to its present form.

This book represents an abridgement (with the addition of some new material) of my *Evolution of the Microscope* (Pergamon Press, 1967, 80*s*.). The abridged version will, it is hoped, reach a wider audience than the more expensive work and perhaps will prove valuable in stimulating the study of the microscope. Research on the history of the microscope, using modern methods, is only just the beginning and as many old instruments have been preserved, much work remains to be done.

Acknowledgements

I WISH to express my gratitude to my friends Mr. G. L'E. Turner of the Museum of the History of Science, Oxford, Mr. Alan Todd of Micro Instruments (Oxford) Ltd. and to Dr. G. A. Meek of Sheffield University for helpful discussion and advice, also for preventing the perpetration of many mistakes. The drawings were prepared by Miss Christine Court, and many of the photographs were specially taken by the photographic staff of the Department of Human Anatomy; I am grateful for their willing assistance.

My thanks are due to the following for permission to use material: Messrs. J. A. Churchill Ltd. for quotation from Carpenter and Dallinger's *The Microscope and its Revelations* and for the use of a figure (Fig. 4.24) from this book; Messrs. AEI Ltd. (Figs. 7.4, 7.6, 7.7, 7.9); Cambridge Instrument Co. Ltd. (Figs. 7.11, 7.12, 7.13a); Messrs. Degenhardt Ltd. (Fig. 6.4a); Studio Edmark, Oxford (Fig. 4.10); Messrs. Leitz (Fig. 6.11); Metals Research Ltd. (Fig. 6.13); the Museum of the History of Science, Oxford (Fig. 4.10); the Royal Microscopical Society (Figs. 1.11, 1.13, 1.14, 2.10, 3.8, 3.10, 4.1, 4.4, 4.9, 4.13, 4.18, 4.19, 4.23, 5.2, 5.3); Professor Ruska (Figs. 7.1a, 7.2); the Science Museum, London (Figs. 2.4, 2.11, 4.8, 4.16); Messrs. Siemens, Germany (Figs. 7.5, 7.8); the Wellcome Trustees (Fig. 4.2).

CHAPTER 1

The First Microscopes

"LOOK AND SEE" is the advice often heard when one is presented with a question of fact. This saying reflects in everyday terms the importance of our eyes in providing information about our surroundings, so that we can adapt smoothly to the demands of our environment and carry out our daily tasks. The eye is probably the most important of our so-called "special senses" and we rely on it to a very large extent. Normally it serves us very well, but occasionally the eye, together with the associated visual areas of the brain, may play remarkable tricks. These are the well known optical illusions. Such visual gathering of information about the surrounding world is, therefore, an important part of our life. It is also an essential part of the scientific method and it is in this particular respect that the human eye is adequate only up to a point; beyond this it fails to help us, solely because the amount of detail which it can provide is severely limited.

From the earliest times up to the end of the sixteenth century, studies of the structure of natural objects had been performed with the unaided eye. Studies of objects at great distances such as the stars, and very minute objects and structural details of larger objects were limited by the capabilities of the human eye. As more and more scientists took up such studies the need for supplementing the human eye became apparent; when the method of combining lenses to obtain an enlarged image was discovered at the end of the sixteenth century the possibilities of augmenting the performance of the human eye were rapidly realized. Instruments which were designed to magnify objects were soon produced and became known as "microscopes". The demands of the natural philosophers — the name by which the scientists of the day were known — stimulated rapid progress in microscopy and this, in its turn, gave immense benefits to the work of the scientists. Before the development of the early microscope is considered in detail, however, it is necessary to digress and consider the performance of the unaided eye, and how this may be supplemented and improved by means of lenses.

The light rays enter the eye through the transparent cornea and are brought to a focus to form a sharp image on the light-sensitive elements of the retina lining the inside of the eyeball. The bending or refraction of the light is caused by the combined action of the corneal/air interface and the crystalline lens which is situated just behind the coloured iris. Variations in the shape of the lens, caused by the pull of a small muscle, enable the eye to form equally sharp images of objects at different distances; this property is known as accommodation, and enables the normal eye to form sharp images of objects which vary in distance from infinity to about ten inches from the eye itself. This latter distance is usually termed the "least distance of distinct vision" or the "conventional visual distance".

It is a well-known fact that the apparent size of any object varies according to its distance from the observer. If we look at a man who is at the far end of a street he seems to be much smaller than when he is only a few yards away, although quite obviously his size has not changed. The change in the apparent size of the man as he walks towards us is due to the fact that the angle which is subtended by light rays from an object at the optical centre or "nodal point" of the eye increases as the object is brought closer to the eye. This is shown diagramatically in Fig. 1.1; here AB represents an object of constant size which is viewed from three positions in turn. When seen from the farthest point, E, the visual angle AEB is small. At an intermediate position, D, the angle is clearly larger, and when the observer is near to AB, the point C, the visual angle, represented now by ACB, is largest of all.

It is thus obvious that the visual angle subtended at the eye by any object depends on its distance, and it follows that if the object is very small, i.e. subtends a very small visual angle, it may be made to appear larger by bringing it closer to the eye. If we think of the printed letters forming a page of newsprint held five feet from the eye, they subtend a very small visual angle and, therefore, the image on the sensitive retina is extremely small; the result is that they cannot be seen distinctly.

The larger black type of the main headlines subtends a larger angle at this same distance and hence the retinal image is larger and the letters can be read. By bringing the paper one foot from the eye the angle subtended by the small print increases and these letters in turn become legible.

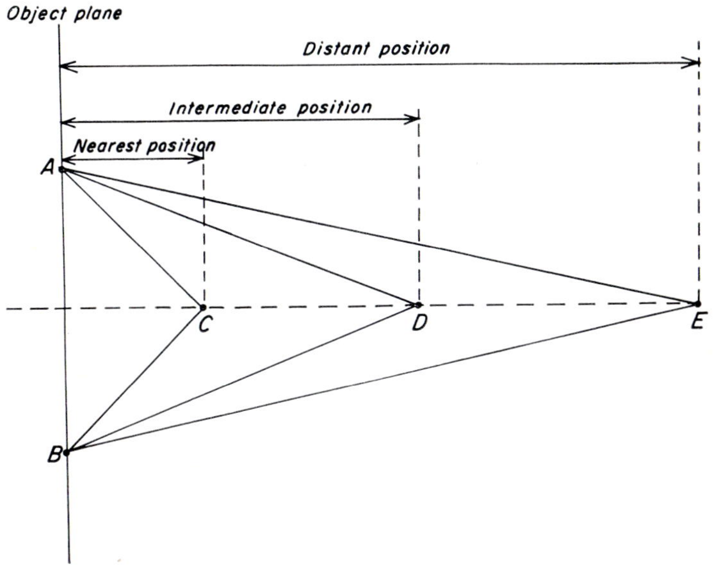

FIG. 1.1. The relationship between visual angle and distance. The angle subtended at the eye of the observer by the object AB increases as one gets nearer to it (represented successively by the positions E, D and C), and the greater the amount of detail which can be resolved in the object.

Experiments and calculation have both shown that when the angle subtended by any object at a given distance falls below one minute of arc it is no longer resolvable by the average human eye and hence its detailed structure would not be visible. In practice there is some variation according to the nature of the object and tests of visual acuity have to be designed to take this into account. A grating of black and white bars observed at the nearest distance of distinct vision (10 inches) by an average person would be seen resolved into its components when there were about 200 lines in the space of one inch.

Since an object must be magnified so that its image subtends at least one minute of arc at the eye, in order for it to be resolved, it might be supposed that this could be achieved simply by bringing the object nearer to the eye, and so increasing its visual angle. The power of accommodation of the eye is limited, however, and objects cannot be placed nearer than the least distance of distinct vision, if they are to be seen in sharp focus.

With very short-sighted individuals, of course, the nearest distance of distinct vision may be as close as 2 inches, so that they can increase the apparent size of fine detail by removing their correcting spectacles and bringing the object very close to the eye.

For the majority of observers the visual angle must be increased in some other way and the microscope is the tool by which this object can be achieved. It must be stressed at this point that we have been talking about *resolution* rather than *visibility*. Resolution, the ability to distinguish two separate points of light or details on an object as separate points, was clearly understood in the seventeenth century by Robert Hooke, although it was treated in relation to astronomy and the observation of double stars, rather than to microscopy. Hooke was the first to conclude that in order to see the stars as double they must be separated by such a distance that they subtend a visual angle of one minute of arc. If they are closer than this, then to most people the two points of light would appear single, so that the visual angle of one minute of arc represents not the limit of visibility but the limit of resolution. It is this property with which we are largely concerned in microscopy. If an image is merely magnified without resolution having been achieved, the result is a blur, which does not convey any more meaningful information and it is then termed "empty magnification". Most people, when confronted with a microscope usually ask "What is its magnification?", when the question which should be asked is about the resolution of the system.

The microscope then serves to increase the visual angle under which the object is seen, thus making it possible, in effect, for the eye to approach very close to the object and still preserve the sharpness of the image thrown on the retina despite the restrictions imposed by the limited powers of accommodation of the eye. Microscopes may increase the visual angle by the use of one or more lenses which enable the eye to form a sharp, spread-out image on the retina, or alternatively by projection of a greatly enlarged real image of the object upon a screen, which is then viewed by the eye in the normal way as though the image were the actual object. Normally when the microscope is mentioned it is an instrument of the first type that is meant, where the image is formed directly in the eye by rays of light from the microscope. The second type — known as a projection microscope — was, however, much used in the eighteenth and early nineteenth centuries in the form of the solar microscope, so called because the illumination was by means of the sun's rays directed into the instrument by means of a special mirror. The projection microscope has recently been revived and is now much used for teaching purposes because the image can be seen easily by several people at once. A modern version of this type of instrument uses a television camera attached to the optical system of the microscope so that the final image may be presented on television monitor screens, which can be seen by large numbers of people.

The increase of visual angle, i.e. the spreading out of the image of the object on the retina of the eye may be achieved either in a single stage by the use of a convex lens (or nowadays a combination of lenses cemented together in order to achieve a more perfect image) or in two stages. In this case the first lens combination, called the objective, furnishes a real enlarged image of the object and this enlarged image is looked at by a further lens system known as the ocular. When only a single lens or combination is used, as in the reading glass or watchmaker's magnifier, it is termed a *simple microscope*; where two or more lens systems are used we have the *compound microscope*.

The action of a simple microscope is shown in Fig. 1.2, in which the visual angle may be compared with that produced by the same size of object placed at the nearest distance of distinct vision shown in Fig. 1.3. In this latter figure the object AB produces an inverted image CD on the retina of the eye. The visual angle is represented by the value α. When a simple convex lens is used as a microscope the object AB is placed closer to the eye than before and the lens interposed between it and the eye. The retinal image C′D′ is now larger than before and

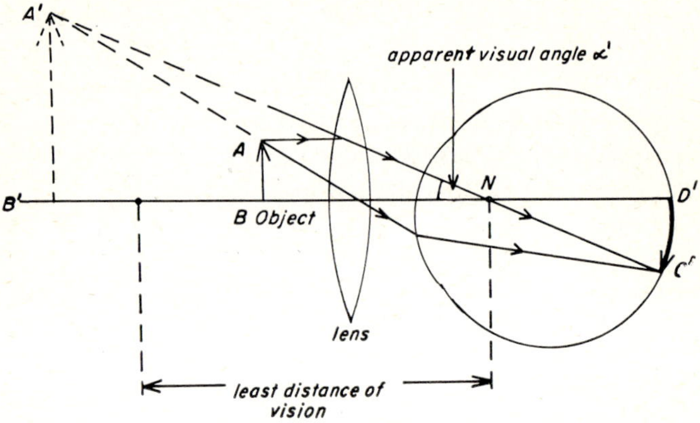

FIG. 1.2. The convex lens acting as a simple microscope. AB is placed nearer to the eye than the least distance of distinct vision and within the focal length of the lens. The virtual image AB appears erect and magnified, and its apparent visual angle (α′) is increased.

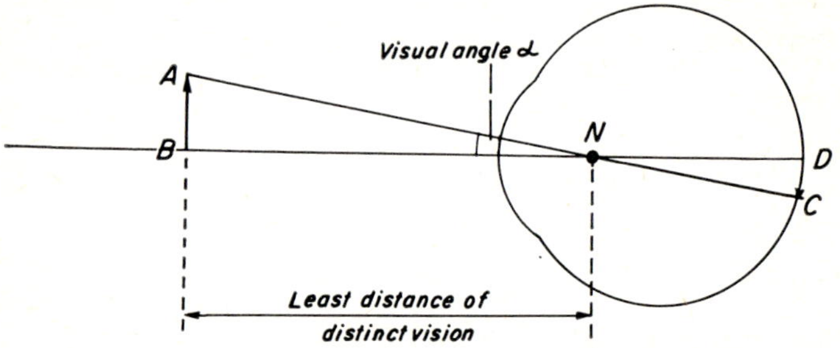

FIG. 1.3. The visual angle (α) subtended by an object AB placed at the least distance of distinct vision is represented diagrammatically in this figure. N represents the nodal point or optical centre of the eye and DC the image of the object on the retina of the eye.

the object appears to be situated at A′B′. This image cannot be projected onto a screen, it does not exist independently of the eye and is said to be "virtual". When a lens is used to throw an image on a screen a picture is formed independently of the eye and in this condition the image is known as a "real" image. From Fig. 1.2 it will be seen that the virtual image produced during the use of a simple microscope is enlarged and the same way up as the object or, as it is termed in optics, "erect". The virtual image of A′B′ now subtends an apparent visual angle of α' which is greater than the true visual angle α produced by the viewing

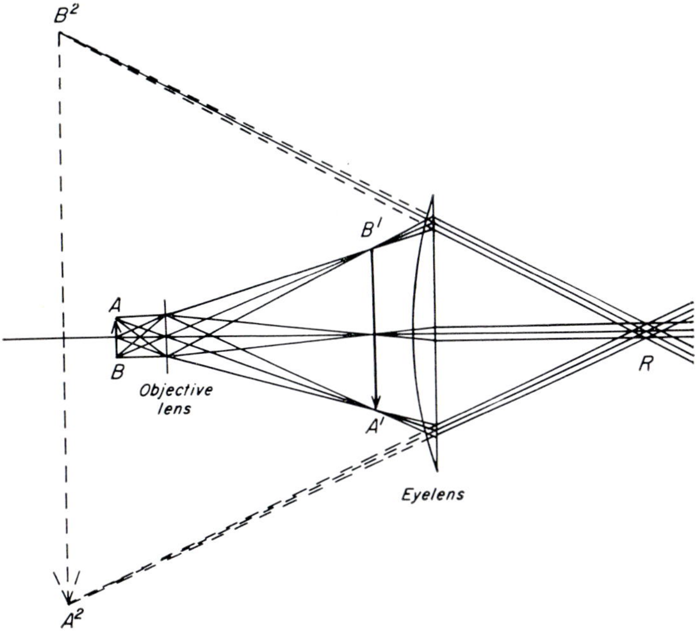

FIG. 1.4. A schematic diagram to show the arrangement of two lenses to form a compound microscope. The real, inverted intermediate image is indicated at A′B′ and the final, virtual image at A^2B^2. R represents the "Ramsden circle" where the rays are brought together after passing through the eyelens.

of the same object without a lens. The virtual image makes it appear to the eye as if the object itself were situated at A′B′ and had that actual size.

The operation of a compound microscope may be shown in a simplified form by a similar type of diagram (Fig. 1.4). Now the object AB is placed just outside the focus of a convex lens or combination of lenses known as the objective. This produces a real, magnified image at A′B′, which is seen to be upside down, or "inverted".

This real image is viewed with the aid of a further convex lens (the eyelens or ocular) so arranged as to produce an erect, virtual image of part of the intermediate image A′B′; this virtual image is represented by A^2B^2 and is still further magnified. As the primary or intermediate image is inverted with respect to the object, it follows that the final virtual image produced by the eyepiece will also be inverted with respect to the object and it will be still further magnified by the action of the eyepiece lens.

It will be seen from Fig. 1.4 that in a microscope constructed with only two lenses, the eyelens must be rather large in order to accept the whole cone of rays from the intermediate image A′B′.

The rays are brought together outside the eyepiece at R, which is found a considerable distance from the eyelens. Point R is called the "Ramsden circle" or "eyepoint" and it is at this point that the eye of the observer must be situated. Considerable practical difficulties exist in manufacturing large lenses which have a correspondingly larger radius of curvature; again, the distance of the eyepoint causes considerable inconvenience to the user for unless the whole cone of rays forming the Ramsden circle enters the pupil of the eye, the whole of the field of view is not obtained. It is not easy to achieve this condition when the eyepoint is some distance from the lens surface.

In order to overcome these difficulties the so-called Huygenian eyepiece, which consists of two lenses arranged as in Fig. 1.5, is now largely used. The larger of the two lenses is called the field lens, the smaller and more powerful being the eye lens. It can be seen from the diagram that the intermediate image A′B′, which in the absence of the field lens would have been produced in the position marked by the dotted lines, is now smaller and formed nearer to the objective. The eyelens, which provides most of the magnifying power of the eyepiece, forms a virtual image of this intermediate image in the usual way. It can be seen that the Ramsden circle is now much nearer to the eyelens (compare Figs. 1.4 and 1.5) and that the diameter of the lenses needed to achieve the same area of field of view is now much smaller. In Fig. 1.5 the size of an equivalent single lens is shown by dotted lines on the right hand side of the diagram.

This eyepiece then has important advantages over a single-lens eyepiece in being easier to construct, as it uses small lenses of great curvature, and in possessing a close eyepoint; even more important, the component lenses can be so calculated that their aberrations cancel out some of the residual errors introduced by the incomplete corrections of the microscope objective.

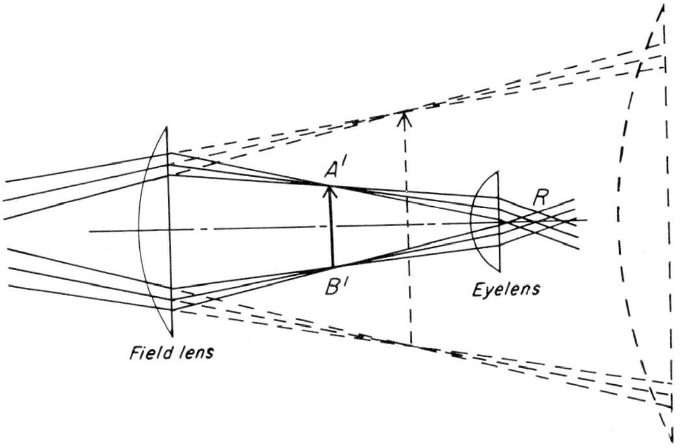

FIG. 1.5. The Huygenian eyepiece. The converging action of the field lens causes the intermediate image (A′B′) to be smaller and situated nearer the objective than would otherwise be the case. The eye lens produces a virtual image of this in the usual way. The size of an equivalent single eyelens is shown dotted in the diagram.

This form of compound microscope having convex lenses in the eyepiece is known as the "Keplerian" type, from the fact that this method of optical construction was first suggested by Kepler in his *Dioptrice* published in 1611. An alternative construction, which substitutes a diverging or concave lens in the eyepiece, is known and is usually called the "Galilean" system, although there is some doubt whether Galileo himself was the inventor.

It seems likely that this system originated in Holland and was taken up by Galileo when he made his telescopes in the early part of the seventeenth century. The Galilean telescope can, by extending the distance between its lenses, also be used as a microscope, and undoubtedly Galileo himself discovered this fact. Under such conditions, however, the Galilean system suffers from the severe drawback of having a very restricted field of view; this, together with the inconveniently large amount of separation required between the lenses, no doubt caused it to fall into disuse for microscope systems.

A passage written in 1614 by Giovanni du Pont, Seigneur de Tarde, quoted in Clay & Court's book on the microscope is interesting in this respect. Du Pont says:

> The tube of the telescope for looking at the stars was not more than two feet in length, but to see objects that are very near, but which we could not distinguish on account of their small size, the tubes must be two or three arms' length.

Galileo used his instrument to look at insects and the same author related that Galileo told him that

> With this tube I have seen flies which look as big as lambs.

As a consequence of the limitations mentioned above, the Galilean type of instrument did not figure in the subsequent development of the microscope, although it has had some limited use in the nineteenth century as a simple dissecting microscope — the so-called "Brücke lens". Its main advantage for this purpose lay in the fact that there must be a considerable distance between the object and the front lens of the instrument which allows room to carry out manipulation and dissection.

There has been much confusion over the claims to be regarded as the inventor of the compound microscope built on the Keplerian pattern. It now seems unlikely that we shall ever be able to settle the matter with certainty, largely owing to the difficulties of interpreting the vague descriptions in the early manuscripts and to the lack of authentic early instruments of known date.

An amusing example of the confusion which can arise from studies of these early writings on this strange new instrument is provided by the *Magia Universalis* of Gaspar Schott published in 1677. Figures in this work show what seem to be giant microscopes, which are nearly the size of a man. Some of these figures are reproduced here as Fig. 1.6. Such instruments would be obviously completely impracticable and historians of the microscope dismissed them as mere flights of fancy. Mayall, however, in his Cantor Lectures on the microscope published in 1888 mentions a suggestion by Frank Crisp, the then Secretary of the Royal Microscopical Society, which seems to explain these strange instruments. The suggestion was simply that the mistake is solely due to the engraver when redrawing the figures. Schott in one case gives the original source of his material and in the other cases it has proved possible to trace them in other contemporary accounts; some are taken from a book *Ars Magnis Lucis et Umbrae* published by Kircher in 1646, whilst illustrations of comparable instruments are found in Traber's *Nervus Opticus* of 1675 (Fig. 1.7).

One tends to estimate the size of the instruments by reference to what is drawn with them and as the engraver instead of drawing simply an eye looking through the small tube (Fig. 1.7a and c) pictured a whole man, one gains the impression that the tube is of an appropriate size (Fig. 1.6 a and c). Again the drawing shown as Fig. 1.6 b would suggest a man looking at a candle flame through an immense convex lens, but on reference to the original source in Kircher (Fig. 1.7 b) one sees that the artist has again converted the representation of an eye into a whole man and that what he has shown as a candle flame was originally meant to represent an insect, impaled on the point of a pin, which is being examined through a lens which must only be a few inches in diameter!

These errors are all the more strange because Schott mentions himself in his text that the tube of one of the instruments

> scarcely exceeds in length and thickness the joint of a finger.

In other cases too he quotes the approximate sizes of the tube of the instrument he is drawing. Perhaps the errors may be due to the still unfamiliar nature of the new instruments and the possibility that the engravings were prepared separately and not seen by Schott. With such hazards abounding in the seventeenth-century literature it is not surprising that doubt still exists as to the early history of the microscope!

In the famous *Dioptrique* published in 1637, Rene Descartes shows two types of simple microscope, one almost certainly the source of

FIG. 1.6. "Giant microscopes" from Gaspar Schott's *Magia Universalis* of 1677. Compare these three figures with the corresponding illustrations in Fig. 1.7, which represent the same type of microscope. A false impression of their scale is gained as a result of the engraver's error.

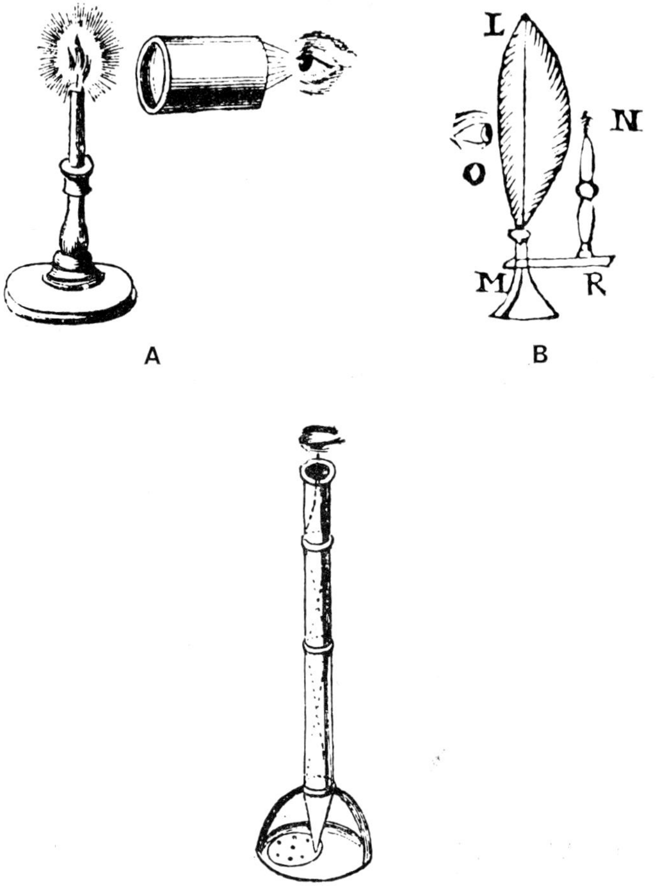

C

FIG. 1.7.

(a) microscope from Traber's *Nervus Opticus* (1675)

(b) microscope from Kircher's *Ars magna lucis* (1646)

(c) Traber's conception of Divini's microscope.

Kircher's figure, the other of a plano-convex lens fixed in a central aperture in what was presumably a metal speculum. This instrument (Fig. 1.8) carried the specimen impaled on a metal spike, labelled G, which held the object in the focal plane of the lens. It seems that the

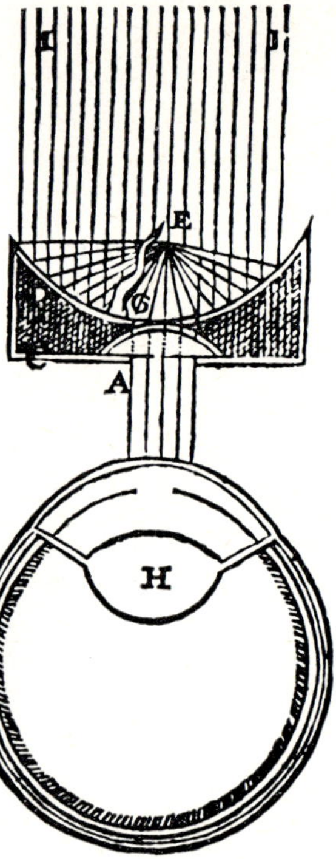

FIG. 1.8. The simple microscope designed by Descartes. The lens is mounted in the centre of a concave mirror which concentrates light upon the specimen impaled upon the spike G.

instrument was designed to be held up to the sun, which was reflected by the mirror so as to illuminate the opaque object. The reflector, which served to illuminate the object, was later carried on the barrel of the objective of a compound microscope, or was fitted surrounding the mounted lens of a simple microscope and was known, after one of its subsequent inventors, as a "Lieberkühn."

In the same book a few pages later, Descartes introduces a figure (reproduced here as Fig. 1.9) of an instrument which is mounted on a stand in a similar manner to a telescope. Again the figure of a man introduced suggests an instrument built on a gigantic scale. There is nothing in the text to indicate the size of the microscope, but again

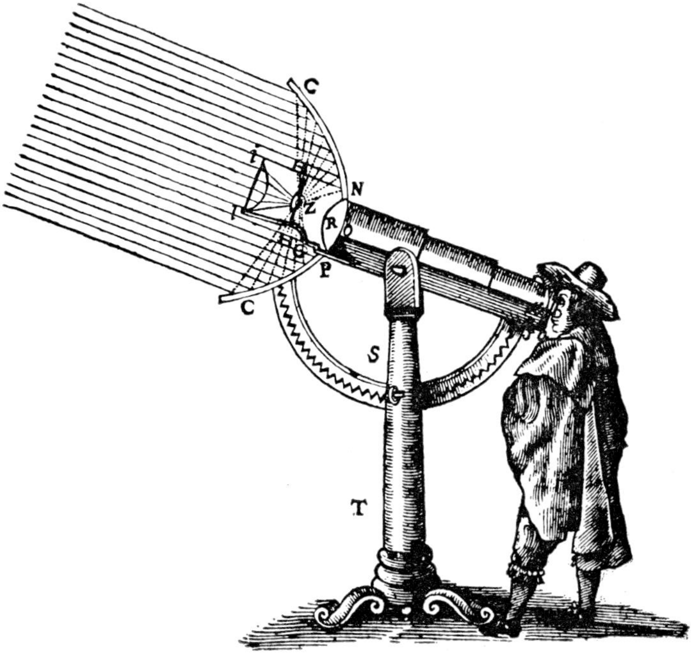

FIG. 1.9. The "giant" microscope of Descartes. Note the "Lieberkühn" type of mirror (CPNC) and the condensing lens (li). The object was placed at Z.

we are probably dealing with a case of draughtsman's licence; the figure which immediately precedes this in the original shows a "perspicillum" mounted on an identical stand, but in this case only an eyeball is drawn at the observer's end. This instrument is obviously a small Galilean telescope and therefore it seems probable that in view of the similarity in mountings, the two instruments would have been of comparable size. The microscope appears to be of the Galilean type, with a bi-convex objective lens, which may not have been of a spherical curvature, and a single concave lens acting as an eyepiece. The object is shown, as in the case of his simple microscope, as being surrounded by a large hyperbolic mirror. Again the idea was to illuminate an opaque object by concentrating the sun's rays onto it. A further very interesting feature, which foreshadows a much later development in illumination technique, is a smaller plano-convex lens (li in the figure) mounted on an arm in such a way as to concentrate a beam of light on a transparent object placed at the focus. This must be one of the first indications of the method which has come to be universal for studying transparent objects with transmitted light. No indications of the focusing methods are given. Although Descartes directs that both instruments should be used with full sunlight, it seems unlikely that either was ever actually constructed. Mayall thought such an instrument would have proved far beyond the rather slender technological resources of Descartes' times. It must be regarded as an intellectual exercise of some ingenuity, but it does furnish one of the earliest printed representations of the combination of lenses in such a way as to form a compound microscope.

As in the case of many useful instruments, it seems probable that the basic idea of how to combine lenses to form a device for magnifying objects occurred independently, at about the same time, to more than one person. Certainly many opticians were active at the end of the sixteenth century, especially in Holland, in the construction of telescopes, so that it is likely that the idea of the microscope may have occurred to several of them independently. Various authorities give the credit to different people and it is difficult to disentangle the rival claims. In all probability the date may be placed within the period 1590–1609 and the credit should go to three spectacle makers of Middelburg in Holland, any or all of whom may have had some part to play in the inception of the microscope.

Hans Janssen, his son Zacharias and Hans Lippershey have all been cited at various times as deserving of the chief credit. From seventeenth-century writing, especially by Wilhelm Boreel, a member of the Dutch diplomatic service, it would seem that the Janssens, both father and on, are chiefly involved. Boreel affirmed in letters (which were later published) that he had known the Janssens personally for a very long time and that Hans Janssen had told him that they were inventors of the microscope. It seems that they had sent one instrument to Prince Maurice of Orange and another to Archduke Albert of Austria. This latter instrument eventually passed into the hands of Cornelius Drebbel and it was then seen by Wilhelm Boreel himself when the latter was envoy to England around 1619.

From his description we learn that it had a tube about eighteen inches long and two inches in diameter, supported by a stand in the form of three brass dolphins. This was placed on an ebony base so that this instrument would only have been suitable for viewing opaque objects by reflected light.

Hans Lippershey's claim to have constructed a microscope is not so well substantiated; it is certain, however, that he did construct a telescope which was unusual at that time in being a binocular instrument. It was made with lenses ground from rock crystal, because of the generally poor quality of the glass available at that time (which was usually greenish in colour and marred by the presence of air bubbles). It was probably for this reason that one of the best known of the seventeenth-century microscopists — Antoni van Leeuwenhoek — occasionally ground his lenses from grains of sand which formed a source of pure quartz.

All these early microscope makers were certainly working quite empirically; as the manufacture of lenses for spectacles was now quite widespread it seems likely that the chance combination of two convex lenses held one above the other in the hand would occur. In some positions they would act as a crude compound microscope from which the actual mounting of the lenses in a tube would be a short step. Whoever took these first steps is not known for certain, but his invention very soon spread widely. From the letter of Wilhelm Boreel, which has already been mentioned, we know that Cornelius Drebbel had a microscope in England in 1619 and other contemporary accounts indicate that the invention reached Rome a few years later.

Harting, one of the great authorities of the last century, certainly believed the compound microscope to have been invented in Middelburg by the Janssens around the year 1590. Rooseboom, however, has recently suggested that this claim cannot be regarded as valid.

Drebbel, if one can believe the testimony of Boreels, actually travelled to Middelburg to purchase a microscope, which he later copied; although he cannot be given any credit for the invention, nevertheless he played an important part in its acceptance as it was instruments made by Drebbel which circulated widely through Europe. Indeed, it was one of his microscopes which Galileo, when visiting Rome in 1624, was called upon to explain to its owner who was unable to make it work. Contemporary letters (later published in the *Journal of the Royal Microscopical Society*) seem to show that this instrument or "occhiale", as it was called, was "a new invention, different from that of Galileo", which showed a flea the size of a locust. Another letter says that "the effect of the occhiale is to show the object upside down and to cause the real motion of the little animal to seem contrary; as for example, if it be going from east to west it will appear to go from west to east". This makes it likely that the instruments of Drebbel were of the Keplerian form with a convex or plano-convex eye-lens.

It has already been suggested that Galileo's only contribution to the microscope was to convert the telescope with a concave eye-lens into a microscope by extending the tube. Once the Dutch Keplerian microscope became known, the shorter, more manageable body-tube length and the greatly increased field of view resulted in all subsequent efforts being channelled into the development of this type. The Dutch thus seemed to be foremost in the field for both telescopes and microscopes at this time, but it must be remembered that the greater part of the credit for the application of these discoveries should go to the Italians. Galileo, in particular, was studying the moon and recording its craters, and the rings of Saturn, whilst in 1625, very shortly after the introduction of Drebbel's microscopes to Rome, we find Francisco Stelluti using it in what must be one of the first micro-anatomical studies on the structure of the honey-bee.

In the museum of Middelburg there is preserved a very old microscope, which is reputed to be an instrument constructed by Zacharias Janssen himself although there is no *direct* evidence to link it with him. It was presented to the museum in 1866 by a private donor, in whose family it had been for several generations. Harting believed this instrument to be a very early compound microscope and it excited much interest in the microscopical world at that time. Since then various copies have been made and a working model may be seen in the Science Museum at South Kensington. The method of construction of this microscope is shown in Fig. 1.10, from which it can be seen that the microscope consists of three tubes, the middle one of which acts as a sleeve to support the outer two, so enabling them to be slid apart. Each of the end tubes contains a lens, that which serves as the objective being plano-convex, whilst the eye lens is bi-convex. The latter lens is loose in the original instrument, but presumably it was once held in place by means of a ring fitted inside the tube. As there is no form of tripod support the instrument was intended for use in the hand.

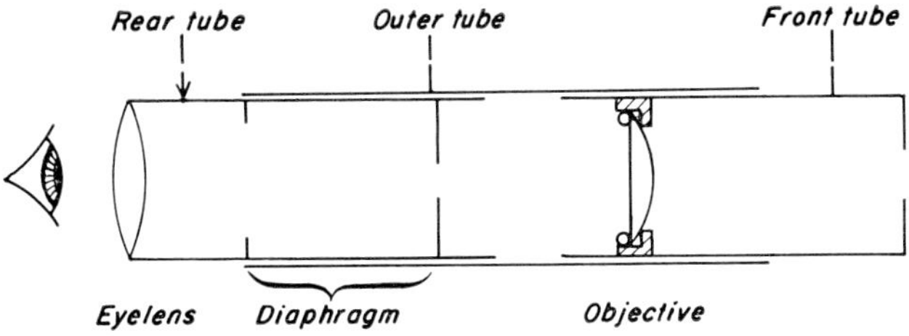

FIG. 1.10. A diagrammatic section through the microscope discovered by Harting in Middelburg and believed to be attributable to Janssen. Note that each lens is mounted in a draw tube which can slide in the outer casing. No stand was provided for this instrument, which was apparently held in the hand whilst in use.

It is obvious that this microscope differs from the Janssen instrument described by Boreel, but as the microscope he was describing was a presentation model the differences in construction may be explained; again the surviving Middelburg instrument may be an early or experimental model intended for some special use or the possibility exists that it is not by Janssen at all.

When we move on to the latter half of the seventeenth century we find that examples of the microscopes of the period are still preserved in museums, though in many cases with the lenses lacking, or in very poor condition. These instruments show very clearly the rather crude methods of construction at first used and how during the course of a very few years the mechanical parts of the instrument began to assume

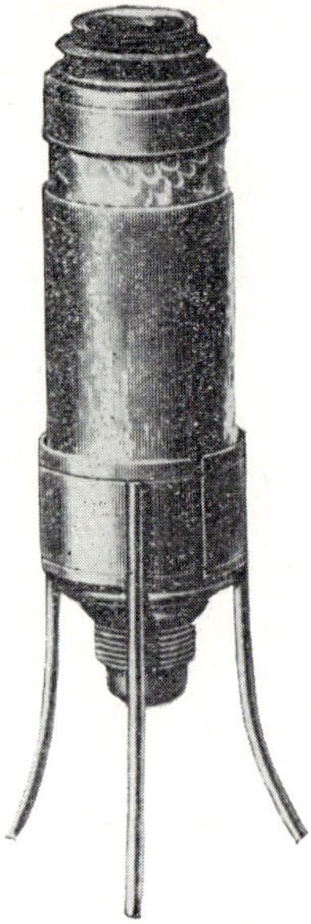

Fig. 1.11. An early compound microscope, probably dating from around 1670, ascribed to Campani. It shows the typical, rather crude methods of construction in vogue at the period.

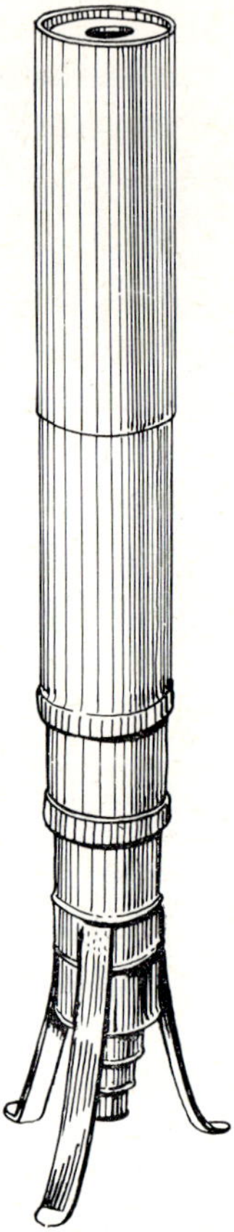

Fig. 1.12. Divini's microscope of 1667. Note the rough tripod mount, the four draw tubes and the absence of any mechanical devices for focusing the microscope.

a fairly standardized form. One early microscope, shown in Fig. 1.11, is attributed by Mayall to the Italian instrument maker and optician Campani, who lived and worked in Bologna at this time. The date of the microscope shown here is doubtful, but it may be thought to represent the state of the art in Italy around the 1670's. As such it is interesting to compare it with the contemporary British microscopes of the same period (see Chapter 2), and note that already there was a considerable divergence in design between the British and the Continental makers, a feature which was to form such a major issue in later nineteenth-century microscopy.

It is obvious that the Continental instrument-makers were content to produce an instrument which was designed solely for the examination of opaque objects which were placed on the table between the tripod feet of the instrument. The tubes are made of cardboard, which was much used in the seventeenth century for optical instrument tube making, and focusing of the lens was effected very simply by sliding the tube within the collar formed by the upper part of the tripod. With the low magnification lenses of rather poor quality in use at this time no very precise focusing mechanism was required and the sliding tube would have served very well and been simple to construct.

A very similar instrument (Fig. 1.12) has been described as the invention of Eustachio Divini around the year 1667.

The similarity of this microscope to that of Campani is apparent when Figs. 1.11 and 1.12 are compared. Both are focused by their sliding paper-covered cardboard tubes; both have a simple type of tripod mount. The lenses of this early Divini microscope do not, however, seem to have survived.

Within a very few years of the construction of the instruments of the sliding-tube pattern, which have just been described, great advances were made in the construction of the microscope stand; in particular the tubes were made of brass, the workmanship was of a much higher standard and the crude tripod was often replaced by elaborate scrolled supporting legs. Focusing no longer was by sliding the barrel inside a collar, but by a screwthread device which altered the distance between the eye-lens and the objective; this device was often combined with a second screwthread for altering the distance of the microscope body from the object.

Two instruments of this pattern are illustrated in Fig. 1.13. For some considerable time these microscopes were attributed to Galileo, but this seems unlikely, since their general method of construction places them at the end of the seventeenth century and Galileo died in 1642. Again no provision is made for the examination of objects by transmitted light; one of the first instruments with this feature is again due to Campani (Fig. 1.14).

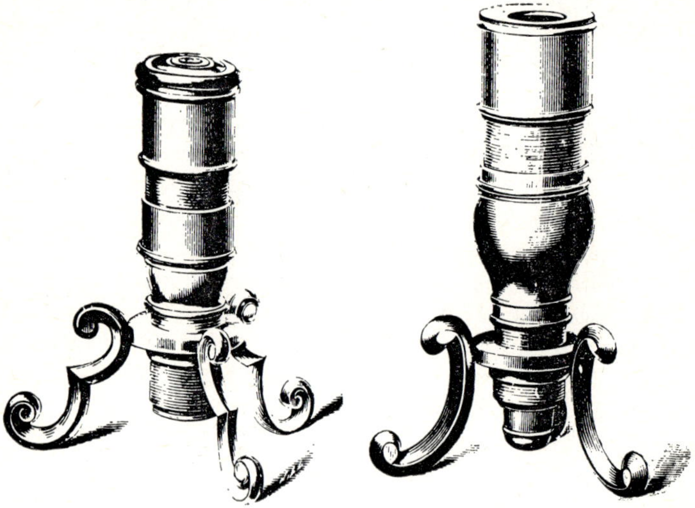

FIG. 1.13. Two brass microscopes from the late seventeenth century, which were wrongly attributed to Galileo. They have no stage plates but are noteworthy for their high standards of mechanical construction. No lenses have survived.

The method of construction of this instrument is very similar to that of the so-called "Galileo microscopes" mentioned above, especially with respect to the focusing movements. Of particular interest, however, are the two plates forming the base of the microscope. The lower plate has two spring clips attached to it, a device which serves to hold it firmly against the upper plate. The object slide could then be sandwiched between the plates, and the object brought into the field of view by sliding the ivory holder until the object was positioned over the central hole which pierced both plates. This arrangement enabled the user to hold the instrument up to the sky or to a candle flame and so examine the specimen by transmitted light; alternatively, the microscope could be placed on an opaque object in order to study its surface.

FIG. 1.14. A later microscope of Campani (*ca.* 1686). Note the double screw thread focusing mechanism and the stage plate with clips and a central hole for the observation of objects by transmitted light.

In the "Descriptio Novi Microscopii, Autore Dn. Josepho Campano, ejusque usus", which appeared in the *Acta Eruditorum* (1686), an account of this microscope is reprinted from the Italian *Giornale de Letterati* of the previous year. This contains a plate which shows the instrument in use both for the examination of a transparent specimen and for observing an opaque object, in this case wounds and scars. At least ten of these later Campani screw-barrel microscopes are known to have survived and they all bear a very remarkable resemblance to one another. It is interesting to note that in this instrument Campani has achieved a radical reduction in size, as his microscope is only about five inches when fully extended and three inches when closed, whereas earlier instruments, such as that of Divini which has been described above, were about sixteen inches or so in height.

One further microscope designer of the late seventeenth century is worth mentioning here, in view of the very advanced design features which he incorporated in his instruments. Bonanni, in his book *Micrographia curiosa* published in 1691 and now something of a rarity, described two types of microscopes. One was a compound microscope of the Campani screw-barrel construction whilst Bonanni's other microscope was of a completely novel conception. It was a horizontal instrument, mounted on its own baseboard (Fig. 1.15). This instrument was designed by Bonanni to fulfil certain features which are of interest as they show many of the factors which are important in the design of instruments today! Translating freely from his original Latin, his aims may be set out as:

1. Convenience in examining the object,
2. A convenient movement for bringing the instrument to the object (i.e. for focusing),
3. The focusing movement should be gradual and occur without danger of losing the object in the field of view, which often happened when the focusing was effected by rotating a screwed tube,
4. The field of view should be equally illuminated,
5. The illumination should remain constant,
6. The instrument should be stable, so that the eye could be removed and when it is replaced the image would still be in focus and the field would still be the same so that the object might be drawn easily.

There can be no doubt that this instrument possessed features in advance of its time. It was probably the first instrument to be given a coarse focusing mechanism worked by a rack and pinion, which is shown in exploded view in Bonanni's drawing (Fig. 1.15). Another unusual feature was the provision of a two-lens condenser mounted onto the lamp house. This concentrated light on to the specimen, as indicated in the figure by dotted lines which Bonanni inserted in order to emphasize this feature which was so novel and important for its success. It is obvious from Bonanni's account that his microscope was in fact actually constructed and made to work, unlike some of the designs which can be found in seventeenth-century books!

As a result of the researches of Bedini it now seems possible to summarize the contribution of the early Italian instrument makers to the development of the microscope. Divini appeared to be responsible for the use of two separate plano-convex lenses as a doublet, united in one mount with their convexities facing each other, and designed to serve as the eyepiece. He also used a combination of draw tubes for focusing and the system of sliding the whole body-tube up or down in its sliding collar in order to vary the distance between the specimen and the microscope objective lens. In some of his later models Divini used a coarse screw thread on the outer part of the lowest body-tube in order to provide more control over the focusing movements.

Campani's developments on the other hand seem to include the invention and the perfection of the screw-barrel device as a means of achieving accurate focusing. This device, which was to survive for many years especially in the simple microscope (see Chapter 3), has often been attributed to another Italian, Tortona, but Campani seems to have the priority. Campani also was responsible for the reduction in size of the microscope, which made it more portable and easier to use and included the slide holder as part of the instrument itself, which helped to bring the illumination system more under control and also allowed the use of a greater variety of illumination methods, especially the use of transmitted light. It is clear that many of the most significant advances in the development of the compound microscope took place in Italy at this period.

The emphasis in this chapter has been upon the mechanical features of these early instruments. It is a great pity that most of the authentic examples which have survived have lost their lenses; of the few which are in working order, such as the little instrument mentioned above by Campani (Fig. 1.14), only isolated measurements of optical

FIG. 1.15. Bonanni's horizontal microscope of 1691. Note the rackwork coarse focus adjustment and the lamp with the substage condenser. The microscope and the lamp are mounted on a single board to ensure that they are in the same relationship at all times.

performance appear to have been made as yet and hence comparison with later instruments cannot satisfactorily be made.

With these instruments the first phase of the development of the compound microscope may be said to be at an end. Subsequent development, especially in England, led to the construction of much larger and more elaborate instruments which will be considered in the next chapter. A tremendous new tool, however, had been placed at the disposal of scientists, especially those studying biology and medicine. Even though these instruments of the seventeenth century, and those which followed for the next hundred years or so, were extremely imperfect optically, they nevertheless stimulated a tremendous interest in the study of minute structure which was to result in remarkable discoveries.

CHAPTER 2

The Compound Microscope in England: 1650-1750

WE HAVE seen that, although the early development of the microscope was largely taking place on the Continent, examples of the compound microscope were certainly sent to England. The letter of William Boreel, quoted in the first Chapter, tells us that Cornelius Drebbel showed him an instrument (almost certainly one made by the Janssens) when Boreel was in England in 1619 as the Dutch envoy. This new instrument, even in its very rudimentary form, must have impressed the English students of the natural sciences, and very naturally they would demand such aids for themselves. It fell to the instrument makers, the opticians and the spectacle makers to satisfy this need.

The microscopes of this period were usually constructed with bodies made of cardboard, covered either with ray skin or, in many cases, with leather finished by embossing or "tooling" with gold leaf pressed in with a metal stamp. Formerly it was thought that the patterns of these stamps were a means of identification of the maker, as each was thought to stick to his own particular patterns. Recent research by G. L'E. Turner in Oxford has now disproved this idea. He has shown that there were three basic tool patterns which were used in successive periods, so that the tooling represents a method of dating an instrument rather than of identifying its maker.

It seems likely that even at this date there was considerable specialization in the manufacture of scientific instruments, and that one or at the most a few workers made all the tubes for microscopes and telescopes for supply to the rest of the trade. Similarly, the lenses and the remainder of the stand could well have been fabricated by "sub-contractors" and the final instrument assembled by the man who actually engraved his name upon it.

Hooke, writing in the Cutlerian Lectures published in 1679, says (with reference to a simple microscope and his design for a compound microscope)

> Both these Microscopes I have directed Mr. Christopher Cock, in Long Acre, how to prepare, that such as will not trouble themselves in the making of them, may know where to be accommodated with such as are good.

This passage suggests that the design and development of the instruments was carried out by gifted workers such as Hooke, whilst the microscopes themselves were constructed by the instrument makers and opticians who were solely artisans who made what they were commissioned to make. It is interesting to contrast this situation with that prevailing in the nineteenth century, when most of the instrumental and optical advances came from the microscope makers themselves and not from the scientists.

Robert Hooke was undoubtedly one of the great personalities of English science of the seventeenth century, and certainly he was one of the first to realize the potentialities of the new invention which had been recently brought to England from the Continent. He was born in 1635 in Freshwater, Isle of Wight, and upon the death of his father he was apprenticed to a portrait painter in London. He soon abandoned this, however, and went to Westminster School and subsequently to Oxford; it is perhaps significant that the originator of the superb microscopical illustrations later to appear in the *Micrographia* obviously had not only artistic talent, but also some formal training in a branch of art which required accurate delineation and observation of detail.

Whilst at the University, Hooke carried out the studies in physics which led to his discovery of the ability of a spiral spring to render the vibrations of a balance wheel isochronous. Hooke had a tremendous breadth of interest, devising amongst other things, a double-barrelled air pump, a spirit level and a marine barometer; his discovery that the elongation of a spring is proportional to the force producing it is still referred to as "Hooke's Law". Hooke became associated with the newly formed Royal Society, and acted as its Secretary and Curator of Experiments. It was during this time that he carried out microscopical studies, and the Royal Society recognizing the importance of this new branch of study, encouraged this; in 1663 he was solicited by the Society to prosecute his microscopical observations in order to publish them and a week later he was ordered to

> bring in at every meeting one microscopical observation, at least.

Hooke faithfully complied with this directive; he showed them the appearance under the microscope of common moss, of the blue mould on leather, the appearance of the edge of a sharp razor and of a point of a needle. He demonstrated various insects, such as the flea, the louse, the gnat and various types of hairs. All these observations, and many others besides, were published in 1665 under the title "*Micrographia: or some physiological description of minute bodies made by magnifying glasses with observations and enquiries thereon*".

Also included in the book are descriptions of his actual microscope, of a lens-grinding machine and detailed discussions of surface tension, refraction and colour. It was a great success and still ranks high today as one of the great masterpieces of microscopical literature. There is no doubt that the publication of the *Micrographia* stimulated a tremendous interest in the microscope, not only among the scientists or "natural philosophers" of the day, but also among the general public, as we may gather from Pepys' diary:

> January 20th 1665 — To my bookseller's, and there took home Hooke's book of Microscopy, a most excellent piece, and of which I am very proud.

The following day's entry concludes:

> Before I went to bed I sat up till two o'clock in my chamber reading Mr. Hooke's Microscopical Observations, the most ingenious book that I ever read in my life.

The great importance of the work of Robert Hooke in microscopy makes it necessary to look at his microscope in some detail and see how it compares with other instruments available at that time.

Hooke's microscope is shown in Fig. 2.1; this instrument is illustrated in the *Micrographia*, but no specimen of this exact type has survived. The instrument bears a very strong relationship to the telescopes of the period, common features being the concave eye-cup which

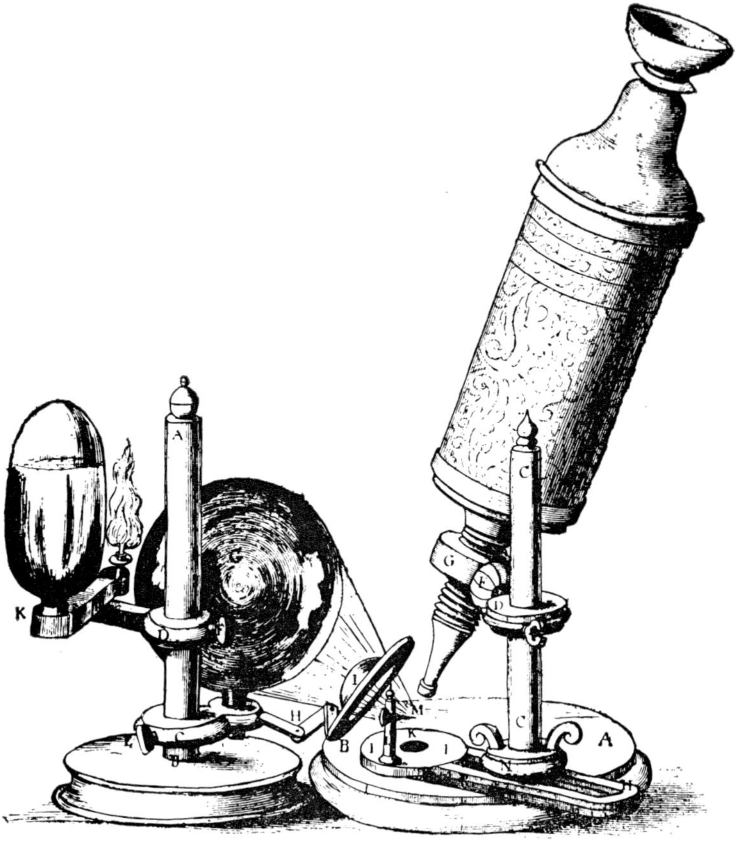

FIG. 2.1. Hooke's own drawing of his microscope, published in the *Micrographia* of 1665. The illuminating apparatus is shown on the left. The nosepiece, with its coarse thread is clearly visible, as well as the mechanical stage.

serves to keep the eye at the correct distance from the eyepiece, the separate draw-tubes sliding one within the other, and the method of mounting on the pillar by means of a ball and socket joint. The body of the instrument, probably constructed of cardboard, after the fashion of the time, was about six inches long when the draw tubes were fully closed. There was a single, bi-convex object glass which was fitted at the lower end of the narrow screwed "snout" attached to the broad body tube. Such an uncorrected objective lens would produce a very poor image. Two aberrations are particularly troublesome in such un-

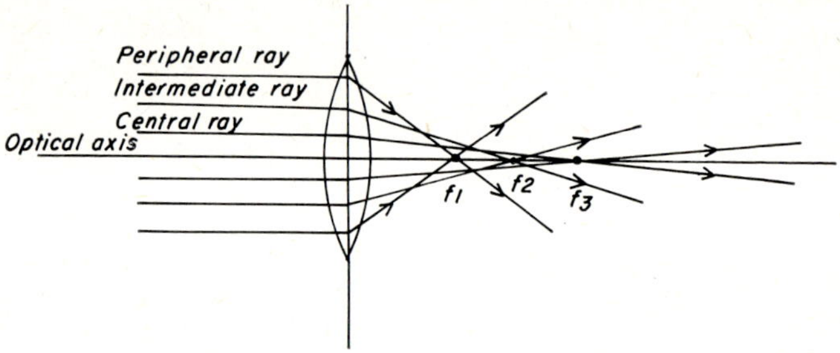

Spherical Aberration

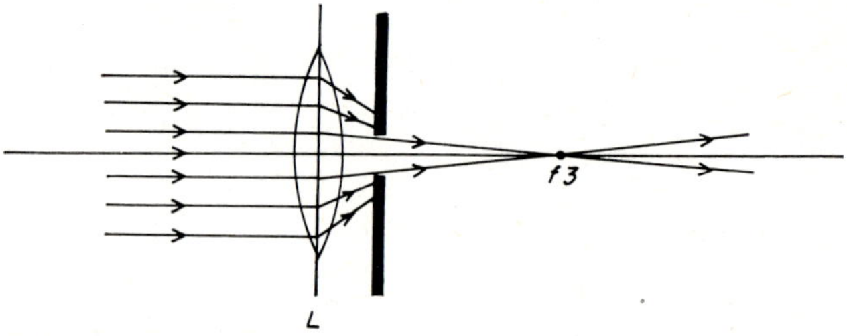

Partial correction of spherical aberration by a stop

FIG. 2.2. The effect of spherical aberration in a convex lens. The peripheral rays are brought to a focus f_1 much closer to the lens than the central rays, which are converged to a focus at f_3. This effect may be minimized by fitting a stop, as shown in the lower part of the figure.

corrected lenses. The first of these faults, termed "spherical aberration", is a defect of spherical lenses arising from the fact that the rays of light passing through the peripheral parts of the lens are more strongly bent towards the axis than those which pass through nearer to the optical axis. This phenomenon results in the focus of the peripheral rays being focused at a point nearer to the lens than the more axial rays, and is illustrated in Fig. 2.2 which shows the focusing of monochromatic light rays by a bi-convex lens. It will be seen that the peripheral rays are focused at a point f_1 which is nearer to the lens than the focus f_2 of the intermediate rays; this in turn is nearer to the lens than f_3 which represents the focus of the rays which passed through the lens close to the optical axis.

In Hooke's time the only way in which it was possible to correct this fault was by the use of a diaphragm with a small central aperture placed behind the lens. This served to stop out the more peripheral rays and prevent them taking any part in the formation of the image; this is shown in the lower part of Fig. 2.2. Such a solution did not prove very satisfactory as it did not completely eliminate the spherical aberration. If the stop were made with an aperture which was small enough to achieve this end, then the amount of light passing was so severely restricted that the brightness of the image was extremely low, making it difficult to see when high magnifications were used. The eventual solution of this problem was not found until the end of the eighteenth or the first years of the nineteenth century, when the development of optical knowledge enabled lens combinations to be calculated which brought the peripheral rays to the same focus as the axial rays and at the same time enabled the other troublesome defect, that of chromatic aberration, to be eliminated.

Chromatic aberration is due to the fact that the different wavelengths which make up white light are bent or refracted to different degrees by the material of the lens. The shortest waves are the most strongly refracted, whilst those with the longer wavelengths are affected least. This means that there will be a series of foci from a beam of white light, extending along the optical axis; the focus for blue light would be nearest to the lens, whilst that for red is furthest away as shown in Fig. 2.3. When such an uncorrected lens is used in an instrument like a microscope the result of chromatic aberration is to surround the image with a series of strongly coloured fringes. This is not only distracting to the eye, but markedly affects the definition of the instrument. Although Hooke was well aware of the nature of chromatic aberration, it was not known as that time how to correct such a defect by combining optical lenses of differing dispersive power and the development of achromatic lenses did not come for another hundred and fifty years. It would be possible to eliminate trouble from chromatic aberration by the use of monochromatic light (that is, light of a single colour) but there is no evidence to suggest that this solution was used by these early workers with the microscope. It seems possible that these early lenses of the seventeenth century suffered far more from spherical aberration and that most of the efforts at correction were directed towards this end. With severe restriction of aperture in an effort to mitigate this trouble, chromatic aberration may have been of less importance.

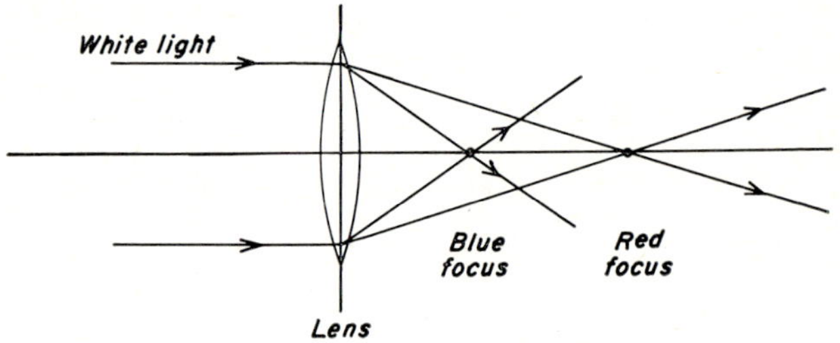

FIG. 2.3. The effect of chromatic aberration in a convex lens. The shorter wavelengths (blue) of the white light are brought to focus closer to the lens than the longer (red) wavelengths.

It must be remembered that if the microscope is a compound one, i.e. one in which the image produced by an objective is examined and further magnified by a second lens acting as an eyepiece, then any aberrations due to the lack of corrections in the objective may well be augmented by the effect of the eyepiece lens. It was this feature, especially, which as we shall see in the next chapter, led many microscopists to persevere with the use of the simple microscope even though the compound microscope possessed some advantages such as a greater working distance and greater comfort in use.

Hooke was almost entirely concerned with the examination of objects by reflected light, a technique much in use today in industrial microscopy and in the examination of materials; the majority of biological microscopy, on the other hand, is carried out with transmitted light, i.e. light which has passed through the specimen. From his experience Hooke found that observations made by natural daylight were unsatisfactory; the light from the sky or sun is very variable and in many cases it proved impossible to complete an examination due to the uncertainties of the climate

> the Weather is so dark and cloudy, that for many dayes together nothing can be viewed.

Hooke found that it was often much more convenient to be able to work at night, so he devised the oil lamp and apparatus to concentrate light on his specimen. This is shown attached to his microscope in Fig. 2.1. In this drawing the lamp is shown at K, the globe of brine (G) served to concentrate the light on the plano-convex lens (I) which in turn could be moved around on its adjustable arm so that the light could be placed where it was needed. Hooke seemed to be pleased with this arrangement, for he wrote

> with the small flame of a Lamp may be cast as great and convenient a light on the Object as it will well endure; and being always constant, and to be had at any time, I found most proper for drawing the representations of those small Objects I had occasion to observe.

A facsimile of this apparatus has recently been built in Oxford by Dr. J. R. Baker and, when tested with the microscope shown in Fig. 2.4, proved very effective.

One of the optical features of Hooke's microscope was the introduction of a lens at the upper end of the body tube, so that it came between the objectives and the eyepiece lens. This third lens or "field lens" has often been thought to have originated with Robert Hooke, but this does not seem to be the case. Clay and Court (the authors of a standard book on the history of early microscopes) believe the field lens must be attributed to Monconys who described it, and tells us that it was made in 1660 by the "son-in-law of Wiselins", who was said to be Campani, although no documentary evidence of this relationship has survived.

There is evidence, from a letter written by Dr. Henry Power (who included a section of microscopical observations in his *Experimental*

Philosophy published in 1663) to Reeves the instrument maker, that the latter was supplying microscopes fitted with a field lens; this would be between 1660 and 1662 and it seems that Power himself was using a microscope with this device in 1661. The field lens in a modern microscope is part of the eyepiece and serves to correct some of the aberrations remaining in the image formed by the objective, but in these early microscopes it must have added still further to the imperfection of the optical system. The only use known for the field lens was to increase the field of view; when Hooke wished to examine fine detail, he invariably removed this lens!

The mechanical form of Robert Hooke's microscope was well in advance of the general practice of the time. All the previous instruments had been of the tripod construction, where the body tube was solidly supported by three legs. This arrangement fulfilled one of the first requirements of a microscope stand, namely that of furnishing a stable mount for the lenses, but the tripod form suffers from the great drawback that the positioning of the lens with respect to the object is very limited. This is not of much consequence with the examination of transparent specimens, for which purpose the lens system must be mounted to move at right angles to the stage holding the object and in the same optical axis as the illuminating system. With the type of microscopy which was predominant in the seventeenth century, i.e. the examination of opaque objects by means of reflected light, it is often very important to be able to incline the microscope tube with respect to the specimen. This feature is provided by the large pillar type of stand, seen in the microscopes of Hooke and in the later Marshall instruments and those of the other seventeenth- and early eighteenth-century makers and it has persisted in a modified form up to the present day.

For focusing the microscope Hooke relied on the screwing up or down of the whole body by means of the rather coarse thread on the "snout" of the body, which worked in the ring labelled G in his picture (Fig. 2.1). This method would seem to have suffered from the rather poor workmanship, which provided a thread of rather coarse pitch which would have the drawback of rather too rapid movements and also probably to very rapid wear. The ring G was in turn attached to the upright supporting pillar C of the microscope by means of a large ball and socket joint F, which was clamped to the pillar by a wing nut passing through the ring D. The object of this complex fitting was to enable the user to vary the height of the body tube and its angle of inclination in order to suit the particular object under examination. This particular method of achieving the inclination, which was essential as microscopes became larger in order that the user could work in comfort, would not have proved very satisfactory in practice as the ball-and-socket joint would have been difficult to make and would have suffered from wear and soon become slack in use. In other instruments made in the next ten years or so (including one reputed to have been used by Hooke, and now in the museum at South Kensington) this ball-and-socket joint has been replaced by a straight rod fixed firmly into the ring which slides on the pillar; the threaded ring into which the objective screws is now fixed to the rod by a joint secured by a clamp screw, so that inclination of the microscope body may still take place but the body can now be fixed much more securely. This instrument, shown in Fig. 2.4, also shows numerous other small refinements, such as possessing a larger pillar, it has only two draw tubes, the eyepiece cup has been abandoned, and the "snout" has a much greater diameter and a finer thread; these latter features would ensure greater rigidity of the body tube and a much more precise method of focusing. It is interesting, however, to note that in both instruments the means which was used to hold and position the object was the same, suggesting that this proved one of the most successful aspects of the original design.

When an object was simply held under the microscope for examination the results must soon have been considered unsatisfactory; not only would the object be magnified but also every minute tremor which is present even in the steadiest hand, so making accurate observations very tiresome. This problem was overcome by use of a feature designed to hold the object and allow of its movement in any desired plane. Such an accessory (now called a mechanical stage, of which that devised by Robert Hooke is probably the first) is found on almost every large research microscope used today. The device used by Hooke was more complex than a typical mechanical stage, being more closely related to the accessory used at the end of the last century for holding and rotating small opaque objects in any desired plane.

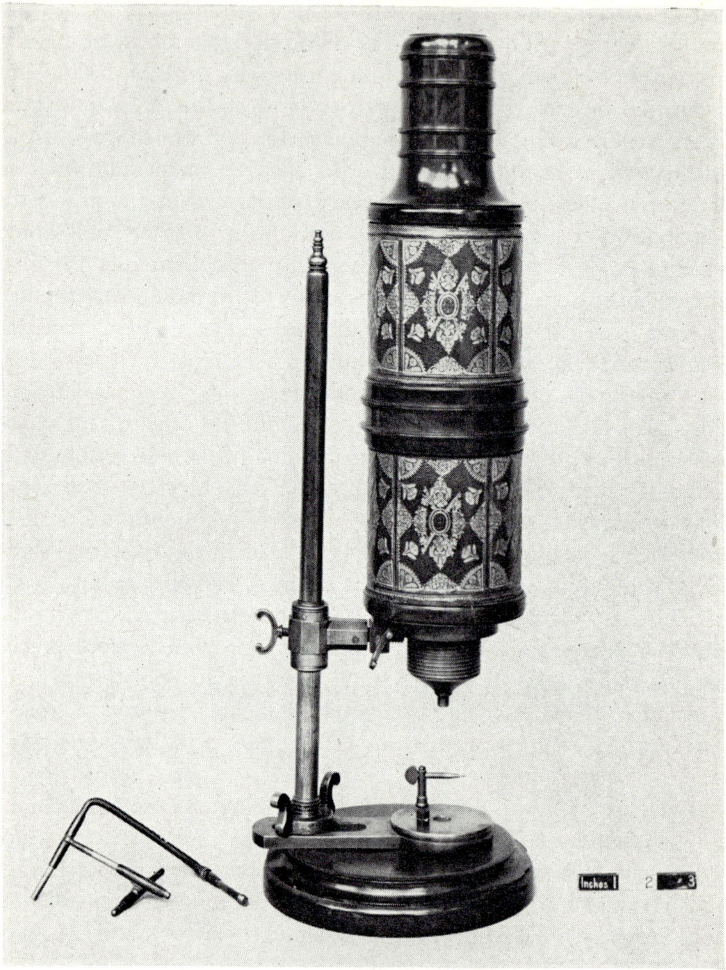

FIG. 2.4. A photograph of a microscope reputed to be made about 1678 for Hooke by Christopher Cock. Note that there are now only two draw tubes, and the nosepiece has a greater diameter and a much finer pitched thread. The workmanship of this instrument is of a very high order.

(*Crown copyright, The Science Museum.*)

The essential feature of any device of this type is that the object is firmly held so that precise movements may be imparted to it by mechanical means. Leeuwenhoek had achieved the same end in a rather similar way with his simple microscope (see Chapter 3), but the form of mechanical stage adopted by Hooke was of a much more elaborate construction. It may be seen in Fig. 2.1 and in detail in Fig. 2.5. The object was attached to or impaled upon an iron pin (M in the figures) which could be rotated; the pin was mounted through a small pillar L, which in its turn was carried at the edge of a round plate (I, I in the figures). This plate was free to rotate around the central

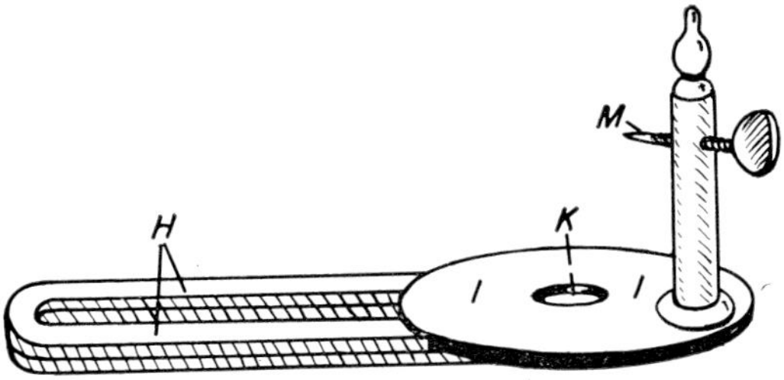

FIG. 2.5. Details of Robert Hooke's mechanical stage. M is the rotatable pin, to which the object may be fixed; the pillar is eccentrically mounted on a disc I, which can rotate about the centre K. This disc in turn is carried by the slotted link H which fits under the microscope pillar. See also Fig. 2.1.

point K, so that the whole provided a means of orientating the object with respect to the optical axis of the microscope. Further flexibility was provided by attaching the rotating plate to the end of a brass slotted link (H in the figure) which was clamped at any position by tightening the wing nut N on the pillar of the microscope. This stage is very carefully designed and evidently proved so successful that a very similar design was adopted by Marshall for use with his "great double microscope" which came into use somewhere around the year 1693.

It was with the instrument which has been described above, mechanically sound but optically rather crude, that Robert Hooke produced his superb observations of common objects. At this time there was no

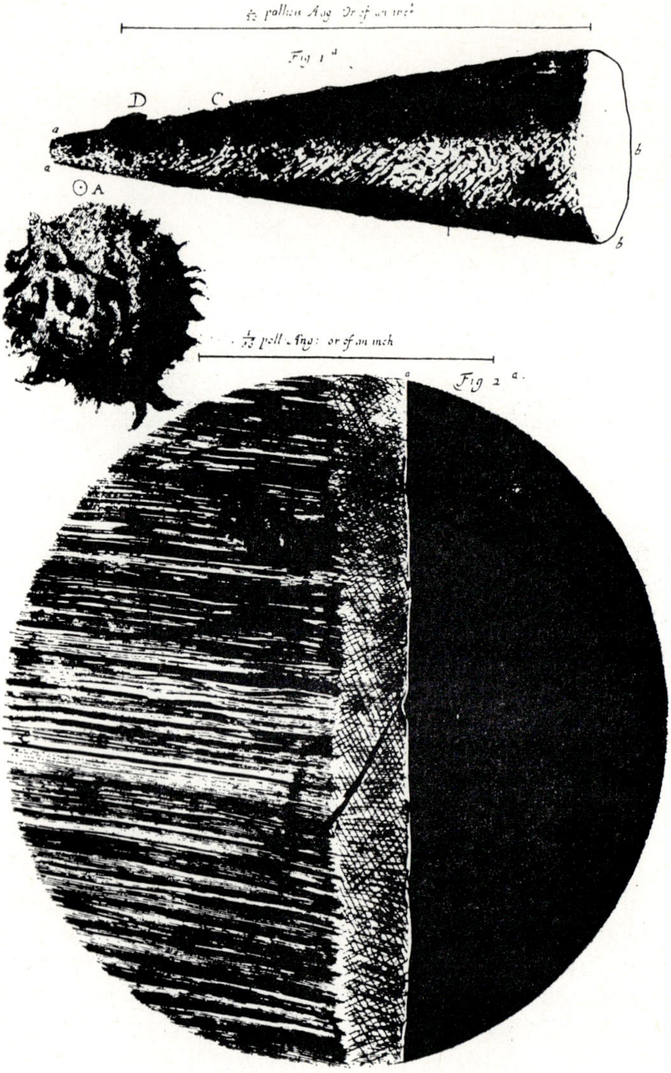

FIG. 2.6. Hooke's drawing of the point of a needle, a printed full stop and the edge of a razor. (Figures 2.6 to 2.9 inclusive are taken from the *Micrographia* of 1665.)

systematic science of microscopy, everything was new and even the most trivial of everyday objects was a source of great wonder when seen under magnifications of up to a hundred diameters. Hooke was a most meticulous observer who was clearly aware that an erroneous

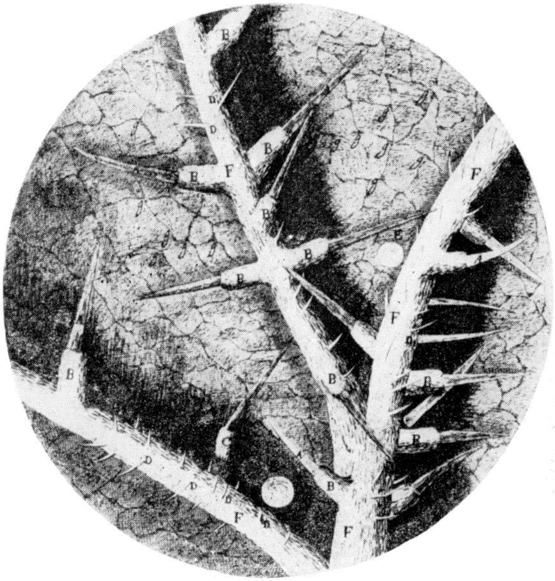

FIG. 2.7. The surface of a nettle leaf.

impression of structure might be gained when studying an object under the microscope. He says that his constant endeavour was to

> first discover the true appearance, and next to make a plain representation of it. This I mention rather, because of these kinds of Objects there is much more difficulty to discover the true shape, than of those visible to the naked eye, the same Object seeming quite differing, in one position to the Light, from what it really is, and may be discovered in another.

He notes the difficulty which one finds in recognizing whether there is a prominence or a depression and of distinguishing between a reflection and a true whiteness.

When Hooke was carrying out his microscopical studies the subject was new and strange and techniques of preparation of objects were rudimentary or non-existent. It is not surprising, therefore, that much of the *Micrographia* is devoted to the obvious way of using the instrument, i.e. looking at the surface detail of an object which was illuminated from above. With this technique the most commonplace objects took on a new form. Figure 2.6 shows his drawing of the edge of a razor and the point of a needle whilst Fig. 2.7 shows the surface of a nettle leaf with the stinging hairs and the outlines of the epidermal cells carefully represented.

The range of observations is very wide; seeds, hair, fabrics, crystals and insects all are faithfully depicted. One of the best known of Hooke's drawings of an insect is reproduced here as Fig. 2.8. The original plate of this flea is truly superb, folding out of the book, and is over 16 inches in length. Perhaps the most significant observation in view of subsequent developments in biology is the illustration of the structure of cork, which is presented as "Observation XVIII: Of the Schematism or Texture of Cork, and of the Cells or Pores of some other such frothy Bodies". In this Section Hooke faithfully illustrates the appearance of the surface of freshly cut cork (Fig. 2.9) and states

> these pores, or cells, were not very deep, but consisted of a great many little Boxes, separated out of one continued long pore, by certain Diaphragms.

Hooke goes on to measure them and found that there were rather more than a thousand in the length of an inch and from this he estimated that there would be

> in a Cubick Inch, above twelve hundred Millions, or 1,259,712,000 a thing most incredible, did not our Microscope assure us of it by ocular demonstration.

He realized that these cells contained air and that this was the reason why cork was so light and floated on water, and why it has such a spongy texture and springiness when compressed. This is the first use in biology of the term "cell", a concept which is now of fundamental importance in the definition of the functional units of living organisms. More than a century was to pass, however, before the realization of their unitary nature was enunciated by Schleiden and Schwann. There

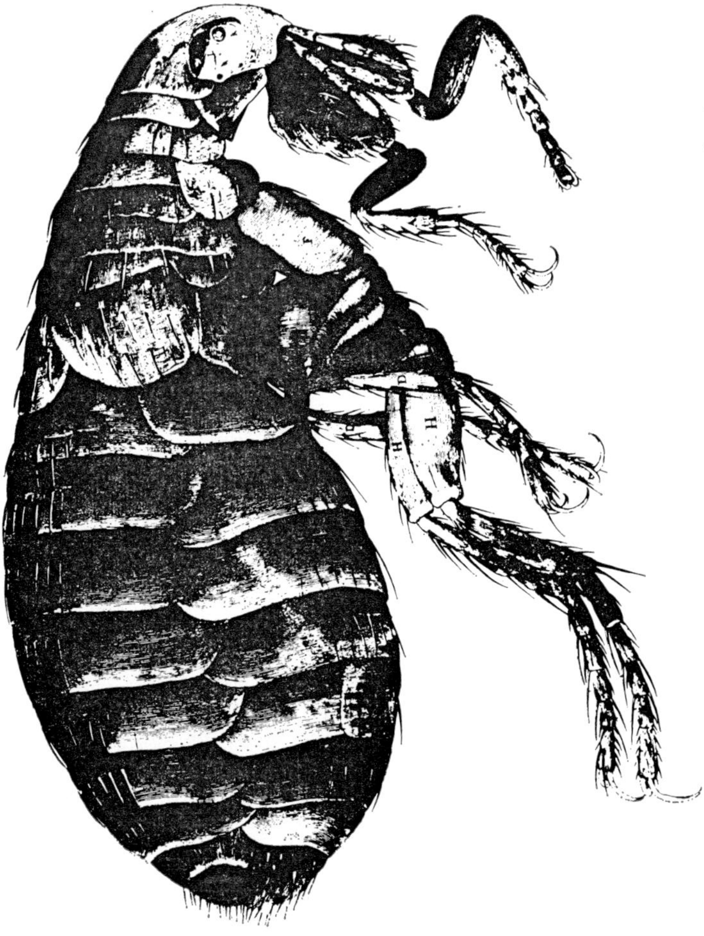

FIG. 2.8. The famous drawing of a flea. The original is over sixteen inches long.

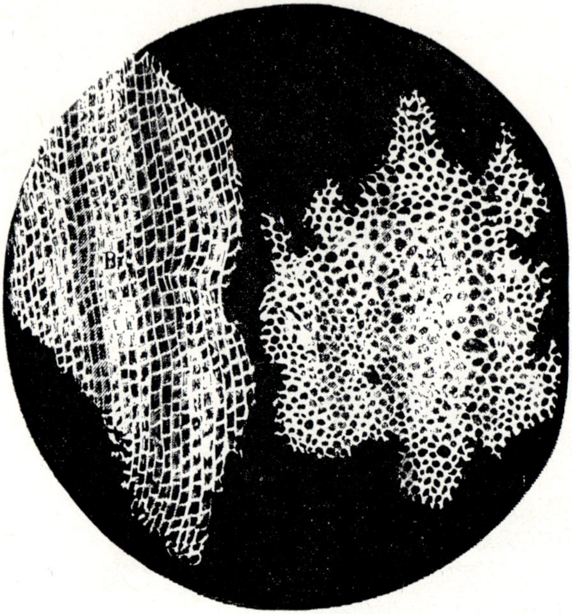

FIG. 2.9 The cut surface of cork.

can be no doubt that Robert Hooke discovered the factual basis of the cell theory; not only did he picture the cell walls of plant cells, but he also saw their contents, the protoplasm. He writes:

> for in several of those Vegetables, whilst green, I have with my microscope, plainly enough discovered these Cells or Pores filled with juices, and by degrees sweating them out.

The time was not ripe for the comprehension of the principles of cellular organization and of the production of new cells by the division of pre-existing cells, so the credit for the cell theory as a biological concept undoubtedly belongs to the later workers. Nevertheless, the genius of Robert Hooke stands revealed by his observations.

Hooke, in addition to preparing superb detailed drawings, also made estimates of the size of the objects which he was examining with his microscope. This may be regarded as perhaps the first example of micrometry or the measurement of the dimensions of minute objects. He did this by the simple expedient of placing a ruler on the base of his microscope, and looking at the divisions on the rule with his unoccupied eye while, at the same time, observing the object down the microscope with the other. The result of this was to create the impression that the magnified image of the object was superimposed upon the ruler so that the *apparent* diameter of the structure as seen with the microscope could be measured; this could then be compared with the actual diameter measured directly. In this way, Hooke was able to estimate the magnification of his instrument.

Hooke was one of the first, and certainly the greatest of the English microscopists, but others were rapidly accepting the challenge posed by the new instrument and were exploiting it, despite its obvious imperfections for biological research. One of Hooke's contemporaries was the Italian Marcello Malpighi; he was born in 1628, near Bologna at which University he graduated as a doctor in 1653. Throughout his life Malpighi was harrassed by professional jealousies and by personal troubles and family squabbles, in spite of which he proved himself to be one of the most original scientists of his time. Much of his work was in the field of physiology, in particular the study of respiration. He published accounts of the lungs and the circulation of the blood through them in the form of letters to Borelli who at the time (1661) was Professor of Science at Pisa. Malpighi described the lungs of frogs which he studied with a small compound microscope of the type shown in Fig. 2.10 from which it is clear that his instrument is very closely akin to those of Campani. He showed that the substance of the lungs consisted of a network of thin-walled alveoli which are connected to all the branches of the trachea or windpipe; he further speculates that the lungs serve to keep the blood flowing and prevent its coagulation. The significant observation is his discovery (by microscopical inspection) of the capillary circulation in the lungs, which connects other arteries to the veins and so completes the pulmonary circuit of the blood. Malpighi made the point that this can be seen in the living animal only with difficulty but he managed to confirm it by looking at the dried inflated lungs of an animal in which the coloration of the blood was preserved. Malpighi himself believed this observation to be important, an impression which has been supported by the view of later workers. It is doubly interesting because it seems that this is one of the first applications of the microscope to the study of living tissues, a branch of microscopy which is now of the utmost importance in biological research.

Malpighi made many other observations with the microscope, using among other techniques the method of injection of an opaque substance into the blood vessels of an animal in order to make them more easily

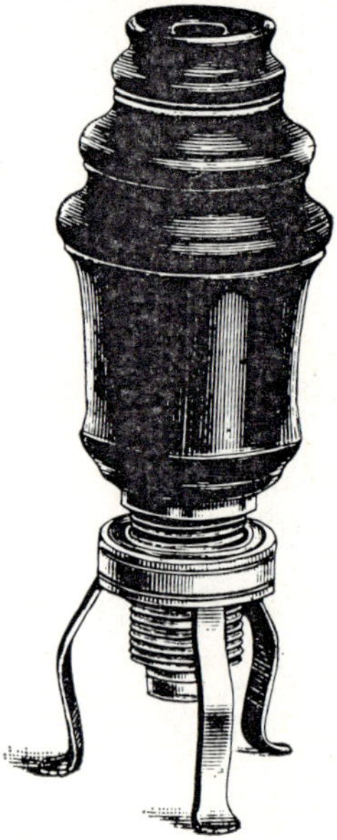

FIG. 2.10. An Italian "vase" microscope of the type used by Malpighi.

visible under the microscope. He established the vascular pattern in the kidney, and in so doing he gave a good description of the microscopic structure of this organ, certain regions of which still bear his name. Further proof of his skill and care as an observer is furnished by his embryological studies, in which he describes the development of the chick. Some of his drawings and descriptions of various organs, such as the heart, could hardly be bettered today. Malpighi also studied insect anatomy, discovering their excretory organs which are now termed "Malpighian tubules" and also he pioneered a new field of vegetable anatomy, in which his microscope showed him the cells or, as he called them, "utriculi" and a great deal of the organization of the tissues of the plant. He showed for the first time the stomata or pores on the under surfaces of the leaves, the vessels with their spiral thickenings in the cell walls and a great many other features which are so familiar to present-day students of plant anatomy.

His compound microscope was more rudimentary than that of his English contemporary, Robert Hooke, but there can be no doubt that as observers they ranked equal.

The tremendous interest in microscopy at the end of the seventeenth and in the early eighteenth centuries was stimulated to a large degree by Hooke's *Micrographia*, but also by the publication by Harvey in 1628 of his discovery of the circulation of the blood. This led to the desire for an instrument which could enable a direct visualization of this interesting process and so most microscopes sold in the eighteenth century were provided with accessories for this specific purpose.

The new instruments stimulated interest, and investigations began to progress rapidly into the techniques for preparing objects for study. They were no longer solely concerned with looking at the surface structure of opaque objects but were beginning to examine biological material by transmitted light, a procedure which often yields much more information. Hooke himself was again in the forefront, for as he says in the Cutlerian lectures with respect to the examination of liquids:

> because the common Pedestal hitherto made use of in Microscopes is generally not so convenient for trials of this nature, I lay those by, and instead thereof I fix into the bottom of the Tube of the Microscope, a cylindrical rod of Brass or Iron. Upon this a little socket is made to slide to and fro; and by means of a pretty stiff spring, will stand fast in any place. This has fastened to it a joynted arm of three or four joynts, and at the end a plate about the bigness of a half-crown, with a hole in the middle of it about three quarters of an inch wide; upon this plate I lay the Muscovy glass, and upon that I spread a very little of the liquor to be examined; then looking against the flame of a Candle or a Lamp, or a small reflection of the Sun from a globular body, all such parts of the liquor as have differing refraction will manifestly appear.

Hooke also realized that in order to see structures in fluids such as milk or blood, they had to be spread as very thin films; he gives detailed directions for preparing films of the right thickness by squeezing drops of liquid between two clear glass plates and even suggests a device with a frame and clamping screws to assist in the spreading. This gadget must surely be the ancestor of the compressorium, which was used so much by the Victorian amateur microscopists in their studies of pond-life! Another of Hooke's very far-sighted comments is that in order to see the structure of muscles and tendons they must be dissociated and examined not in air but

> in a liquor, such as water, or a very clear oyl.

Here is an observation, which forms the basis of modern mounting techniques where the difference in refractive indices between the object and the mounting medium is known and controlled in order to enhance or suppress contrasts and which plays a large part in certain measurement methods which are used today with phase-contrast microscopy (see Chapter 6). Again, Hooke working purely with empirical methods had hit upon one of the basic principles of microscopy. It is clear that this branch of science had made enormous strides in the space of ten years and the foundations of the modern study of histology (the detailed structure of the tissues of the body) had been laid by a few isolated observers, led by Hooke and Malpighi.

Another extensive worker with the microscope around this time was Nehemiah Grew; indeed, it seems probable that he actually used the instrument commissioned by the Royal Society from Christopher Cock and which may have also been used by Hooke. Grew was educated as a physician, in Cambridge and Leyden where he graduated in 1671. He returned to London and applied himself to medical practice and at the same time carried out scientific research. Like Hooke, Grew was elected a Fellow of the Royal Society and in due course became its Secretary. Nehemiah Grew was much more of a specialist in his studies than the other workers we have described, concentrating on the anatomy of plants. His descriptions of the stem, the root, the fruit and the seed all testify very clearly to his tremendous enthusiasm for the subject and to his passion for detail. His researches were published in full in the *Anatomy of Plants* which appeared in 1682 and established him, together with Malpighi, as the founder of the systematic study of plant anatomy. It seems strange to us, looking back on these tremendous discoveries in the field of plant structure, that they were regarded as little more than curiosities of Nature, and no attempt was made to follow up or exploit them. It may be that the chief interests were devoted to animal anatomy, also being revolutionized by the microscope, because of its much greater relevance to the medical sciences.

Hooke, Grew and Malpighi, together with that other genius of the microscope — Antoni van Leeuwenhoek (who worked, however, entirely with the simple microscope) — form the great quartet of workers who established the microscope in the seventeenth century as a tool of scientific investigation; by their prolific writings and observations they stimulated the curiosity of men and excited interest in the study of the minute details of the structure of common objects. At this time when everything revealed by the microscope was new and exciting, the manifest optical imperfections of the instruments did not prove such great drawbacks to microscopical discovery as might be imagined. The aberrations only prove significant when high powers are used and a tremendous field lay open which was well within the limited capabilities of these early instruments. It was only in the following century, when the need for higher and higher powers arose, that the aberrations of the lenses began to cause uncertainty in the interpretation of the images and led to suspicion of discoveries made with the aid of the microscope. This suspicion reached such a pitch that critically-minded people often refused to use or accept as valid any discoveries made with the aid of the microscope. We shall see later (Chapter 5) how such scepticism was only overcome with the introduction of achromatic, corrected lenses, and an understanding in the Victorian era of the scientific principles of microscopical resolution.

Towards the end of the seventeenth century, John Marshall began to make microscopes and working at the sign of "The Archimedes and Two Pairs of Golden Spectacles" he produced around the year 1693 what may be regarded as the natural successor to Hooke's instrument. One of the first figures of Marshall's "Great Double Microscope" (so called because of its size, and to distinguish it from single-lens microscopes) appears in the first volume of John Harris's *Lexicon Technicum* 1704, together with an account of the method of its use.

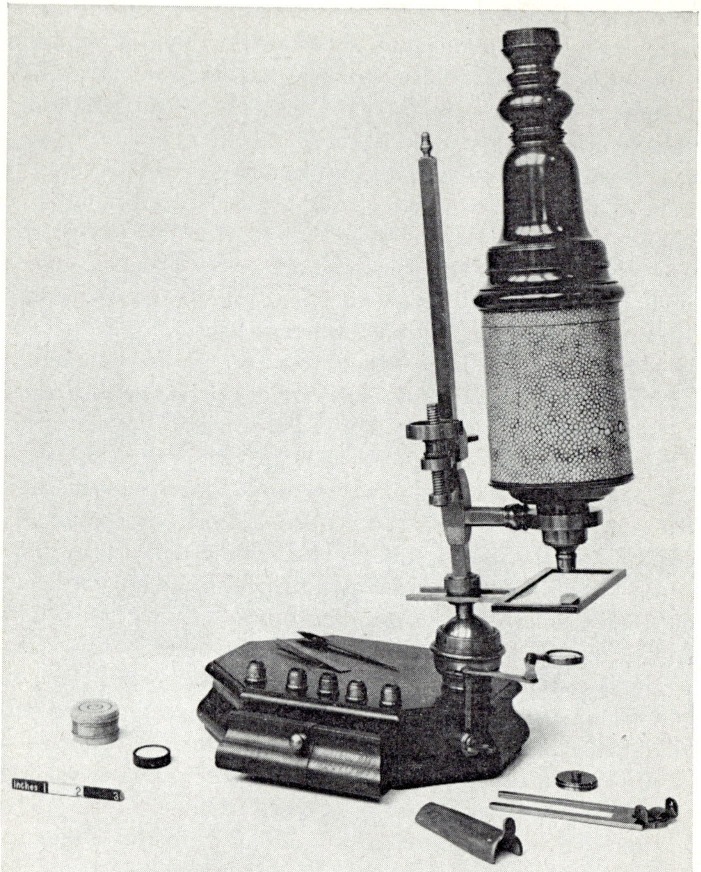

FIG. 2.11. A photograph of John Marshall's "Great Double Microscope". Note the substantial construction of the instrument. Five extra objective lenses are displayed on the box foot together with a pair of forceps and the stage forceps, whilst in the foreground on the right are seen the lead "coffin" for holding down a fish in order to see the circulation of the blood, and the attachment for holding the stage forceps.

(*Crown copyright, The Science Museum.*)

Fortunately at least fifteen of the original instruments are still in existence so that the features of this microscope can easily be studied. A typical example of Marshall's instruments is shown in Fig. 2.11 from which it will be seen that the general form of the microscope very closely resembles that of Hooke. The body of the microscope is carried on a pillar of a square section, about twelve inches long, which is in turn mounted by means of a large ball and socket joint to an octagonal wooden base. In this base, which was very heavily weighted with lead at the end opposite to the pillar, was mounted a drawer which served to carry the spare objectives and the accessories. In this instrument, as in the later Hooke microscope, each of the separate objective lenses was mounted in its own cell which could be screwed onto the nosepiece, whereas in the early Hooke microscope it seems that the various powers of the objective lens were not supplied with separate cells, but that one cell was made to hold any one of the lenses.

The large ball and socket joint, which served to incline the microscope body and allow it to be fixed at any angle up to an inclination of about 45° from the vertical or to allow the body of the instrument to be swivelled right round so as to overhang the box foot, was placed at the bottom of the pillar, an important difference from the Hooke instrument. This meant that, as the specimen stage was carried on the same pillar, the body could be inclined to give a convenient angle for working, without altering the relationship of the objective to the specimen. In this respect, too, Marshall's instrument differed from its predecessors. Most of the contemporary drawings of the Marshall microscope in fact show the body swivelled right round, the weighted box foot keeping the instrument in balance, so that the transparent glass stage could be used for transillumination of a fish's tail with light from a candle placed on a stool or on the floor (Fig. 2.12); a small plano-convex lens fastened on an arm and clamped beneath the object-stage served to concentrate the light onto the object.

Most of these microscopes were supplied with six different objective lenses. Studies on the lenses of several Marshall microscopes have recently been published by the author (see Bibliography); measurements show that Marshall provided an extensive series of lenses, with magnifying powers ranging from about $\times 2$ for the lowest up to about $\times 100$ for the highest. The quality of the image provided by these latter lenses, however, was extremely poor. It seems likely that the selection of the actual lenses to be fitted to any particular microscope would have been made from this series by the purchaser according to the task for which the microscope was required.

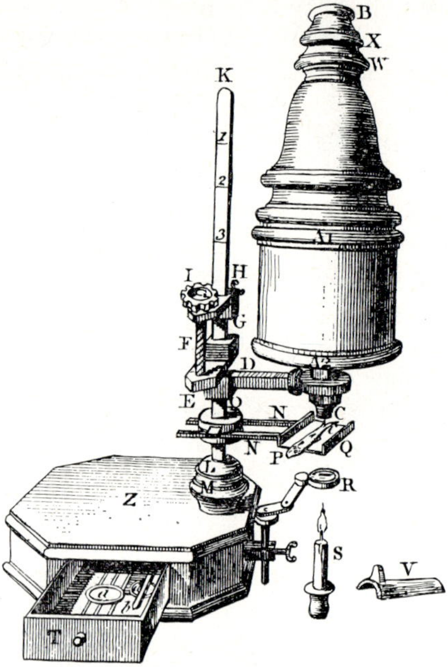

FIG. 2.12. The Marshall microscope in use for examining the circulation of the blood. From a late eighteenth-century encyclopaedia.

One of the characteristic features of the modern microscope, namely the mirror used to reflect the light into the substage, has not yet been mentioned as occurring in any of these instruments. It certainly was not yet in general use. Although one or two undoubtedly genuine Marshall microscopes with a mirror are extant it seems likely that this represents a later addition. As signed instruments made by Marshall in the later years of his life do not possess a mirror, all such modifications were probably carried out after Culpeper had brought it into general use. There is no doubt that the Marshall microscope was to some extent inconvenient to use when set up for examining specimens by transmitted light and the Culpeper microscope (which will be described in Chapter 4) was to a large extent adapted to remedy this fault. In addition, the Culpeper instruments would be much easier to manufacture and therefore would be much cheaper than large instruments of the Marshall type.

Contemporary records give us some idea of the price of these early large microscopes. The Royal Society paid Christopher Cock £8. 6s. for the instrument which he supplied to them, whilst Samuel Pepys tells us that what was probably a very similar microscope from Reeves cost him £5. 10s. It is probable that a Marshall instrument would cost about the same, so that in terms of the money values at that time a microscope was not to be considered cheap. In the *Lexicon Technicum* John Harris wrote about Marshall's microscope in these glowing terms:

> and I take his Double Microscope here described, in all respects to be the most useful, handy and ready Instrument of this kind. I have had Mellen's Glasses, and seen Leeuwenhoek's and Campani's but I would sooner have the Double Microscope than any of them, and the Price is much easier.

If the purchase of a microscope constituted a very considerable financial outlay in those days then perhaps it is not very surprising that there was no tremendous follow-up of the pioneering studies of the early workers with the compound microscope. All the great microscopists of the seventeenth century were somewhat eccentric in their habits and did not seek to train pupils or found a school of followers.

Also we must remember that many people of that time considered their complex and unusual studies of the microscopical structure of things as almost sacrilegious, in seeking to pry into the secrets which God had obviously not intended to be revealed to the eyes of men. As a consequence there was no great use of the microscope for scientific purposes and the eighteenth century was to a large extent a barren period of microscopy. The instrument which had proved so powerful a tool in the hands of Hooke, Malpighi, Grew and Leeuwenhoek was to a large extent demoted to the status of a toy to entertain the idle rich.

Such work as was carried out was done not with large instruments on the Marshall pattern, but with either a Culpeper type of microscope, which in many ways represented a retrograde step in design, or with a simple microscope.

CHAPTER 3

Simple or Single-lens Microscopes

WE HAVE seen in the previous chapters that much interest was shown in the development of optical devices for assisting the human eye; some of the various types of compound microscope which were devised for this purpose have already been described.

At the same time the property of a single plano- or bi-convex lens to furnish a magnified image in the manner indicated in Fig. 1.3 must have been noticed. When such lenses were fastened into a mount or cell, so that the eye was forced to look through the axis of the lens, the first "single microscope", as Hooke called it, was born. At first such lenses would have had a low magnifying power, of the order of a few diameters only, and these instruments were in common use for the examination of insects, especially fleas, hence the name often given to such magnifiers of "vitra pulicaria" or "flea-glasses".

One of the earliest pictures of a flea-glass is to be found in Kircher's book *Ars Magnis Lucis et Umbrae*, published in 1646. This instrument is shown here in Fig. 3.1. The lens was a small sphere of glass mounted in the end of the tube AB, which was closed at the end A by a flat piece of glass. The object to be studied was attached to this glass at C and viewed by pointing the whole at the candle flame, labelled D in the illustration.

At a later period several different types of flea-glass were made, with lenses of different magnifying powers and often they were sold as a set, all being supplied in a small case of leather or wood covered with fish skin. Some of these are still extant and a set in the National Museum of the History of Science at Leyden contains seven different flea-glasses, each of which consists of two cylinders about an inch long and slightly less in diameter, which fit closely one inside the other. One cylinder carries the lens plate, the other the object and hence focusing can be carried out by sliding the inner tube a greater or lesser distance into the outer cylinder.

Another early form of the simple microscope, little more than a mounted magnifying glass, is illustrated and described in Zahn's *Oculus Artificialis* of 1685 (shown here as Fig. 3.2a). The lens was fixed in an elaborate wooden holder on a stand which was transfixed by a rod. This in turn carried the arm bearing the specimen. This latter, often an insect or a part of a flower, was focused by sliding the

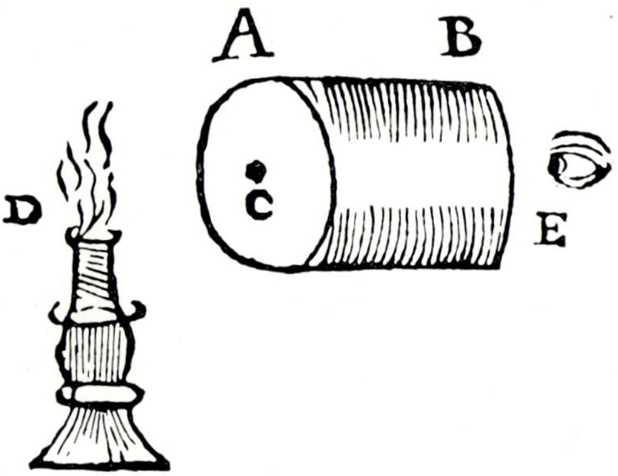

FIG. 3.1. A "flea-glass", as pictured by Kircher 1646. D represents a candle flame, C the object under examination mounted on a plane glass. AB is the tube of the simple microscope, whilst E represents the eye of the observer.

rod through the upright pillar, so varying the distance between the lens and the specimen. This particular example, according to Zahn, had a bi-convex lens with a magnifying power of fourteen diameters. Several similar instruments have survived in our museums. Some of them, undoubtedly intended for botanical work in the field, have the table stand replaced by a turned wooden handle, whilst others instead of having the specimen mounted upon a rod which was slid through the stand, had a simple spring device to serve as the object holder (Fig. 3.2b). This is probably the simplest possible form of controlling the specimen/lens distance, as the object was moved nearer to or away from the lens by pressure of the hand. Such a crude device would only serve for examination at very low powers.

Experience with such simple magnifying lenses no doubt led serious workers to attempt to increase their power. In practice, in

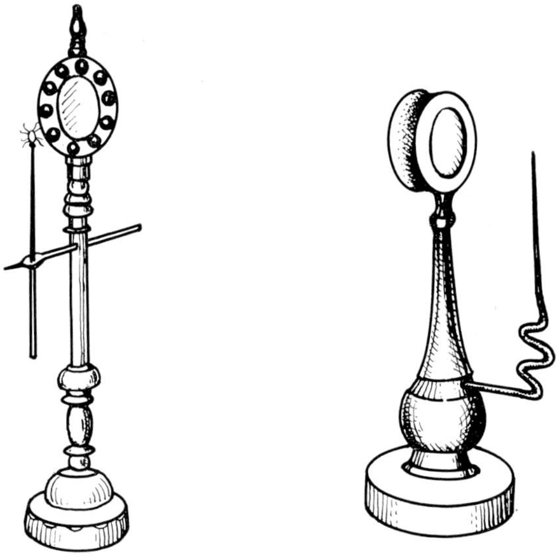

FIG. 3.2.

(a) A simple microscope mounted on a stand. The object is impaled upon the rod carried in the cross arm. After Zahn, *Oculus Artificialis.*

(b) A similar instrument in which the object is carried on a spring arm so that focusing may be easily carried out.

order to increase the power of a simple lens, its radius of curvature must be increased. This, as shown in Fig. 3.3, leads from A, a simple plano-convex lens, via B, a bi-convex lens of the same radius, through stage C of increased curvature and power to the condition drawn at D, where the maximum curvature has been reached for that diameter, resulting in a spherical lens. Further increase in power can only be obtained by reducing the diameter, as shown at E, a solution which has the drawback of demanding both the eye and the object to be placed even closer to the lens. The construction of such small lenses was not easy in the seventeenth and early eighteenth centuries, as lens making was an art still in its infancy. The great advantage of the simple microscope in providing an image relatively free from chromatic aberration soon led to the use of small lenses produced by melting glass capillary rods in a flame and to greater and greater efforts at grinding and polishing lenses. Most of the simple microscopes which have

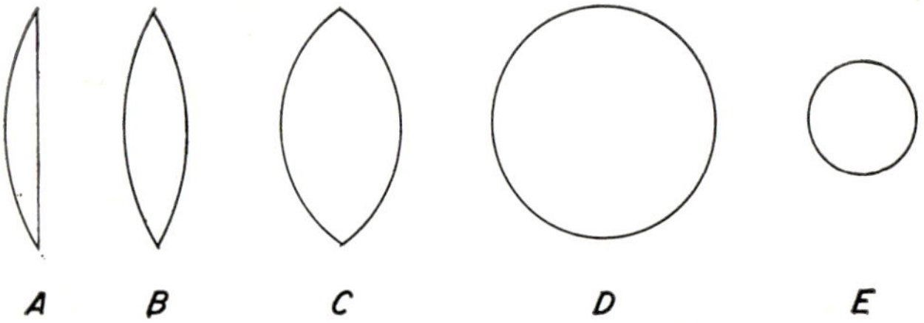

FIG. 3.3. The increase in power of a simple lens; (a) represents a plano-convex lens. A biconvex lens (b) of the same curvature would have a greater power. Further increase is possible by increasing the curvature (c) until the lens is spherical (d). Any further increase is then only possible by reducing the radius of the sphere, as shown at (e).

survived possess lenses made of melted or "blown" glass, as this type was much easier to construct.

Hooke, although he preferred to use the compound microscope, described in the last chapter, was well aware that advantages were to be found in the use of the simple microscope:

> in truth they do make the object appear much more clear and distinct, and magnifie as much as double Microscopes: nay, to those whose eyes can well endure it, 'tis possible with a single Microscope to make discoveries much better than with a double one, because the colours which do much disturb the clear Vision in double Microscopes is clearly avoided and prevented in the single.

The other great English microscopist of this period, Nehemiah Grew, writing in 1681 in his *Catalogue of the Rarities belonging to the Royal*

Society agreed with this view, and gives us some valuable information about simple microscopes in England at this time. Contrasting the compound with the simple microscope, he says:

> The advantage of one with more Glasses, is that it takes in a bigger Object, or a greater part of it. Of one with a single Glass, that it shows the Object clearer. So that to have a distinct representation of it, 'tis convenient to make use of both. Of the latter kind, I have seen several made by Mr. John Malling in this City, not only with melted, but with Ground-Glasses so very small, that one of these Ground-Glasses being weighed in the Assay-Scales in the Tower, was found not above the fourscorth part of a Grain. The Diametre or Chord 1/25th part of an inch. Another, so small, that those Scales were not nice enough to weigh it. The Chord hereof to that of the former, is as two to three. These are the clearest and best that ever I saw.

The reference here is almost certainly to Mellen, an instrument maker who had a very high reputation at this time for the excellence of his microscopes. It is obvious, too, that although he produced "blown" glasses, Mellen also undertook the grinding of lenses at this time. Blown glasses would be the easier of the two to make, but it must be remembered that in the seventeenth century even the production of simple melted glass spheres would not be the easy, reliable operation that it is today.

One further advantage of the ground lens over the "blown" spherical lens is that the image produced by the latter is only sharp over a very small portion of the centre of the field, i.e. the usable field of view is very small. In biological work this is often undesirable and the only remedy is to use ground and polished lenses. In order to emphasize this difference van der Star photographed a scale through a Musschenbroek sphere and through a Leeuwenhoek ground lens of nearly comparable resolution and magnification. With the spherical lens the sharp area of the image produced at a distance of ten inches from the lens was about half-an-inch in diameter, whereas with the Leeuwenhoek lens the area of the field which was acceptable had a diameter of two inches. This lens, with a magnifying power of 275× proved superior to a Zeiss doublet lens of the second half of the nineteenth century, which had a power of only 155× and a sharp area of the projected image at ten inches from the lens of only one inch! Leeuwenhoek must certainly be acknowledged as one of the master lens-makers of all time.

Before passing on to a detailed consideration of the microscopes of Leeuwenhoek and some of the results which he achieved with them, it is appropriate at this point to stress that the simple microscope, although possessed of an undoubted optical superiority, was not much used for investigations involving a high degree of magnification. There are several reasons for this. First, the single lens, because of the short focal length needed for high magnifications, has to be brought very close to the eye. Hooke, in particular, found this an insuperable drawback.

> I have found the use of them offensive to my eye, and to have much strained and weakened the sight, which was the reason why I omitted to make use of them.

If the lens is spherical then, as mentioned above, the field of view is rather restricted. This proved a serious drawback to the use of such lenses for serious biological work, especially as the devices for moving the specimen relative to the lens were extremely rudimentary in the seventeenth and eighteenth centuries. Again, one must remember that at this time the greater proportion of the microscopical examinations were carried out by means of incident illumination, that is, light reflected back from the specimen through the lens system of the microscope. With single lenses of high power, not only did the lens have to be close to the eye but also it had to be equally close to the object; the "working distance", as it is termed by microscopists, was inconveniently short and this interfered with the satisfactory illumination of the specimens. This objection, of course, does not apply to transparent objects or to thin sections of specimens which are viewed by transmitted light, the light actually passing through the specimen before reaching the lens.

Finally, we have seen in Fig. 3.3, that once the lens has been made spherical, reduction of its diameter, giving smaller and smaller lenses, is the only way of increasing its curvature and hence the power. A very small lens, however, will only transmit a minute pencil of light rays, so that the pupil of the eye, instead of being filled with light, as would be the case with the compound microscope, receives only a very small amount of light and so the image produced by a simple microscope of high power lacks brilliancy.

In spite of these drawbacks, however, the simple microscope survived. Although there was very little original microscopical work in the eighteenth century, such as there was of lasting value proved to be carried out on a single-lens microscope. Even today, although compound microscopes have been brought to a high degree of perfection, uses can still be found for the single-lens microscope.

It is evident from the fact that several eighteenth-century works on optics include detailed directions for constructing glass spheres for simple microscopes that most workers would expect to construct their own instruments; the most noted maker and user of simple microscopes at this time was no exception to this, although he preferred to grind his lenses rather than use the simple "blown" sphere. Antoni van Leeuwenhoek was born in 1632 at Delft, where he lived and carried on business as a draper.

He was eventually appointed to the position of Chamberlain to the Sheriffs of the town, a position which he retained until his death; evidently he was held in high regard by his fellow citizens for in 1679 he received the further appointment of "wine-gauger", an office entailing the assay of all wines and spirits entering the town and the checking of the vessels in which they were contained. Leeuwenhoek presents the figure of a prosperous businessman, active in civic service but in 1673 he began to come into prominence in the world of scholarship as a result of the publication of a letter from him in the *Philosophical Transactions* of the Royal Society. This letter dealt with some observations on the sting, mouth parts, and the eye of the bee and of the louse.

A few years previously the *Philosophical Transactions* had published an account (which had been translated from the Italian *Giornale dei Letterati*) of Divini's new microscope. In 1673, Reinier van der Graf, a fellow countryman of Leeuwenhoek, wrote to the Secretary of the Royal Society, Henry Oldenburg, and told him of Leeuwenhoek's achievements.

> A certain most ingenious person here, named Leeuwenhoek, has devised microscopes which far surpass those which we have hitherto seen, manufactured by Eustachio Divini and others.

Van der Graf enclosed Leeuwenhoek's letter which Oldenburg translated and published in the *Philosophical Transactions*. The observations contained in this letter obviously pleased the Fellows of the Royal Society and Oldenburg was instructed to write to Leeuwenhoek asking for any more observations which he might make; as a result numerous letters, containing many remarkable microscopical discoveries, were sent by Leeuwenhoek to the Royal Society throughout the next fifty years.

Leeuwenhoek became accepted as an international authority on microscopical matters, but he remained in Delft working away on his own. In 1680 he was accorded the honour of Fellowship of the Royal Society, which he esteemed very highly. All his research and studies were reported by him in his native Dutch (for Leeuwenhoek did not know Latin, in which most of the scientific communication was carried on at that time) in letters to the Royal Society and to fellow scientists all over Europe. As Leeuwenhoek's fame increased he was plagued with numerous visits from people, including Royalty, who wished to see at first hand the marvels which he was describing with the aid of his microscopes. He regarded these visits as a regrettable interruption in his work, in which he remained active until very shortly before his death in 1723 at the age of 91.

Leeuwenhoek was largely reticent as to his methods of microscopy and the nature of his instruments. Contemporary writings, such as letters from people who were fortunate in obtaining an interview with Leeuwenhoek, often stress this point. It seems that he had one type of microscope which served as a demonstration model for display to his visitors, but he never showed off his best instruments. There are several references in Leeuwenhoek's letters to this subject; he says that his "best microscopes" he kept for himself alone and in a well-known letter written in 1678 he says:

> My method for seeing the very smallest animalcules and little eels I do not impart to others, nor yet that for seeing many animalcules at once, but I keep that for myself alone.

Apparently Leeuwenhoek used to make all his own equipment and lenses; as it seems that he used each microscope he made for only one or possibly two objects, it follows that he must have made a very large number during the course of his long researches in microscopy. One estimate places the number of his microscopes at over five hundred at the time of his death. Of these, twenty-six were bequeathed to the Royal Society and the rest were sold by auction after the death of his daughter Maria. Unfortunately, of all this number only nine authentic examples are known to have survived, six of which are now still in Holland. The Royal Society bequest was preserved for many years and the instruments were described in some detail by Martin Folkes who was Vice-President of the Royal Society in the year of Leeuwenhoek's death. Later still, in 1740, Henry Baker re-examined them and provided us with a list of their focal lengths, magnifications and the objects mounted for examination in each microscope. These instruments remained in the possession of the Royal Society until about a hundred years ago, when they apparently vanished without trace from the Society's collection.

From examination of the surviving instruments and from the early descriptions it has proved possible to assess the optical performance of some of his microscopes and to study their mechanical construction. The lenses were always ground and polished to the bi-convex form, an operation at which Leeuwenhoek became extremely skilful. One example, measured by van Cittert, has a thickness of just over one millimetre and a radius of curvature of 0·75 millimetre. The finished lenses were mounted between two plates of metal each containing a small hole; the plates were riveted together to hold the lens firmly in position between the two holes. The objects were fixed upon a point, which was then moved into the focus of the lens by a system of screws, which can be seen in Fig. 3.4. This anticipated by many years the moveable stage focusing which reappeared in the eighteenth century and is used today on many research microscopes. All Leeuwenhoek's microscopes are small, the total length of the metal plate, which served as a handle as well as a lens mount, being usually under 2 inches and the width usually about 1 inch.

One is immediately struck by the mechanical crudity and imperfections of these microscopes, the metal plates being poorly finished and the screw threads ill-made. The objects were mounted on the pin and when this was rotated by means of the small knob attached to it some variation both in placing of the specimen and in its height with relation to the optical axis of the lens would be introduced. The height of the specimen was roughly set by means of the long vertical screw which passed through the right-angled bracket attached to the lower end of the metal lens plate. There was a short thumb screw which

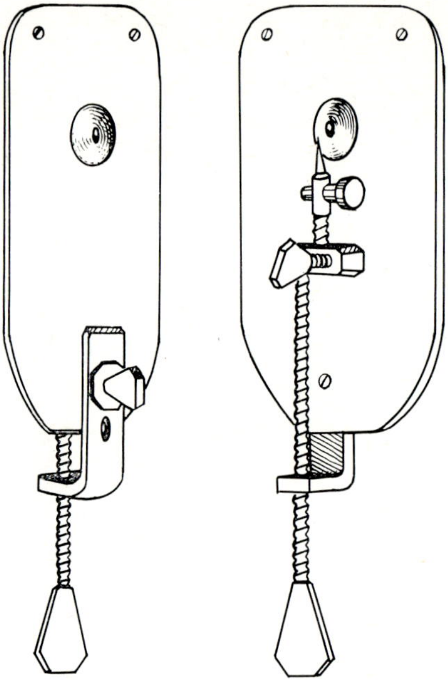

FIG. 3.4. A drawing of both front and back views of a typical microscope constructed by Leeuwenhoek. The pin which carries the object can be clearly seen, as well as the screw threads which serve to adjust its position. The whole microscope was about two inches in length.

passed horizontally through one end of the object support block and which came to bear upon the lens plate. By manipulation of this screw the object pin would have been brought closer to the lens or carried away from it, so effecting the focusing movement.

Leeuwenhoek's lenses were of surprisingly good quality. Van der Star has measured the resolving power and the magnification of several examples. Two of these, now in the Leyden Museum, magnified 79× and 126× respectively and had a resolving power of 3·5 microns. They were characterized, as previously noted, by possessing very large fields of view over which the image was in sharp focus. The instrument in the possession of the Utrecht museum seems to be exceptional. It has a very high magnifying power (275×) and even though the lens is very badly scratched it has an excellent resolution which very closely approaches the theoretical maximum (one micron) for this type of lens. When this lens was new it must have been a magnificent tribute to Leeuwenhoek's skill as a lens maker. It is a characteristic of all his ground lenses — he probably never used "blown" spheres on account of the difficulty of getting satisfactory glass and because of the small field of a spherical lens — that they all have exceptionally large fields of view. The resolution of the Utrecht lens is so good that it is possible that the lens could have been used for the famous observations on bacteria, although some authorities, notably Schierbeek, believe that Leeuwenhoek must have used a lens with a magnification of nearly five hundred times.

This type of microscope, as constructed by Leeuwenhoek, must have been very difficult to use. The working distance is very short, that of the Utrecht instrument measured above has been found to be 0·5 millimetre. The observer's eye would have to be placed so close to the lens that they must have been very uncomfortable to use. Although mechanically primitive and difficult to use, these instruments were optically superior to the lenses made by many of the leading opticians of the day. Furthermore, these microscopes had a rudimentary form of control over the movements of the specimen and, perhaps most important of all, they could be used to study a specimen by transmitted light. This type of illumination which is often the most valuable for microscopy of biological material was effected with the specimen mounted in a short length of capillary glass tubing or spread on a thin piece of blown glass or talc. What does not appear clear is the nature of the illumination, for there does not seem to have been any device for concentrating the light onto the specimen. Leeuwenhoek may have used bright daylight or the flame of a candle as light sources; certainly use of the former is suggested by a passage in a letter written in 1674 where he says:

> I would venture to recommend that, when one of these sections has been brought upon the pin of a microscope, you then hold the microscope towards the open sky, within doors, and out of the sunshine, as though you had a telescope and were trying to look at the stars in the sky through it.

In using such a microscope, in addition to any problems caused by the lack of resolution in the lens, or by insufficient illumination, we must remember that great difficulties in seeing the specimen would be caused by the lack of contrast, due to the transparent nature of living material. There is some evidence that Leeuwenhoek was aware of this particular problem, for Henry Baker says in 1740 that Leeuwenhoek's skill was also shown

> in the contrivance of the Apertures of his Glasses, which, when the Object was transparent, he made exceeding small, since much Light in that Case would be prejudicial; But, when the Object itself was dark, he inlarged the Aperture, to give it all possible Advantage of the Light.

This would certainly help to improve visibility, for reducing the aperture would enhance the diffraction effects around the edges of a transparent object and so contribute to its visibility.

Even though considerable obscurity still surrounds Leeuwenhoek's techniques of microscopy there can be no doubt of the value of his observations, particularly in the fields of bacteriology and protozoology which he initiated. The first account of the free-living, or as Leeuwenhoek describes them "the very little animalcules" is to be found in a letter of 7th September, 1674; the study was continued and the famous letter to the Royal Society in which they are described at length is dated 9th October, 1676. It is a very long letter and is not possible to quote here, but the interested reader is referred to Clifford Dobell's translation in his biography of Leeuwenhoek which makes interesting reading.

In addition to studying and describing free-living protozoa and bacteria, Leeuwenhoek also described the vorticellids, the rotifers and the fresh-water polyp (now called *Hydra*) upon the roots of the duckweed; with his description of *Hydra* he stimulated interest in this organism which lasted throughout the eighteenth century and indirectly influenced the future design of microscopes, as the requirements of workers studying the free-living fresh water coelenterates led to the development in the mid-eighteenth century of the so-called "aquatic" microscope.

Leeuwenhoek's interests did not stop at the inhabitants of water; following Malpighi, he studied the circulation of the blood, and the structure of many of the organs of the body, such as the striated muscle. To this latter end he proved well in advance of his time in the field of preparative techniques, attempting to improve the contrast of some of his material by staining it in a solution infused from saffron. He also investigated very thoroughly the life cycle of ants, aphids and mussels, and with his rather primitive microscopical apparatus provided a tremendous mass of information on these topics; Leeuwenhoek also studied plant structure and other botanical topics. Unfortunately, lack of formal training often resulted in his observations being set down completely without order or apparent purpose, but this does not in any way detract from his achievements. It might well be thought remarkable that all these different spheres of observation could be attempted with such simple home-made equipment, but as van Cittert remarks "these discoveries were made, not in spite of, but thanks to, the simple microscope".

One of Leeuwenhoek's best known contemporaries was also a fellow countryman and a microscopist. Jan Swammerdam had a short career as a naturalist, from 1663 to about 1675 but during these few years he worked with frenzied activity. Unlike Leeuwenhoek he had a formal education in science but his great work in this field, the *Biblia Naturae*, was not published during his lifetime. During the period of his life in which he was a microscopist, Swammerdam carried out a great deal of microdissection of insects, especially the honey bee; he pioneered techniques of dissecting the parts under water and of injecting minute vessels with wax or mercury to assist him in following out their course. Swammerdam's observations were carried out with a simple type of dissecting microscope which was probably made for him by Samuel Musschenbroek of Leyden.

Later, Samuel's young brother Johann also entered upon the business of an instrument maker and achieved great fame as a maker of microscopes and also of air-pumps. Several of Johann Musschenbroek's instruments have survived and they can be readily identified by his trade mark of an oriental lamp and crossed keys (taken from the Leyden coat-of-arms), which he placed on all his products. Johann Musschenbroek devised two types of simple microscopes, one for high-power and one for low-power work such as dissecting. The general form of the construction of the high-power instrument can be seen from Fig. 3.5; these instruments were probably intended to use glass spheres as lenses but the lenses have not been preserved in those microscopes which have been studied and documented.

The objects for study were intended to be fixed onto one or other of a number of rods of differing form, terminating in a single or a double spike, a ring, and so on. These specimen holders are themselves held in a hollow rod which fits into the stem of the instrument. Two screws serve to move the rod around and so position the specimen in front of the lens. The actual mount for the lens is a metal plate, which slides into a holder attached to the handle of the microscope by means of a U-shaped rod. As the lens holder has a crude form of dovetail slideway for the lens plate it is to be supposed that several lenses of different powers were provided, each in its own little metal plate. Some instruments, although not by Musschenbroek himself, have survived with their lenses mounted in this manner.

Over the lens mount was fitted a hollow box which served to exclude light from the sides. A hole was drilled in it exactly opposite the lens position and in front of this, as can be seen in Fig. 3.5, a sector diaphragm was placed. This latter contained a series of holes of different sizes, any one of which could be brought in front of the fixed hole. This device seems to be the first instance of a diaphragm being used to control the amount of incident light falling on the specimen. A similar device is found in modern microscopes although today it serves to regulate the aperture of the illuminating cone of light and not as a means of regulating the intensity of the light or as a means of increasing contrast. As mentioned before, it seems likely that in these early single-lens microscopes restriction of aperture, either by carefully choosing the size of the lens aperture as Leeuwenhoek did, or by the use of a crude diaphragm by Musschenbroek served as a valuable means of obtaining more contrast in transparent or semi-transparent specimens which were viewed by transmitted light.

Focusing in Musschenbroek's high-power microscope was carried out by the action of one of the wing nuts which served to move the

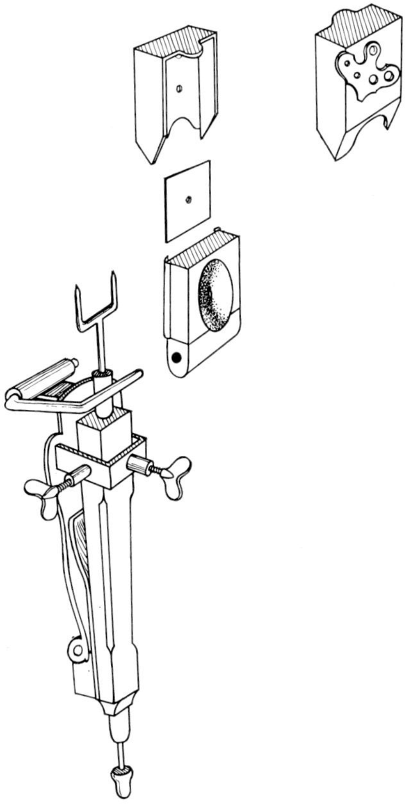

FIG. 3.5. An exploded view of the high-power microscope of Musschenbroek. The handle and specimen carrier together with the focusing screw and object adjusting screw are at the lower left. The lens (mounted in a rectangular metal plate) slides in a wooden holder which fits onto the horizontal bar attached to the handle. A metal cover, which also clips onto the holder is shown at the top of the diagram. This cover serves to exclude unwanted light from falling on the object, and carries a quadrant with different sizes of aperture which can be seen in the front view (upper right of the illustration).

plate carrying the lens either towards or away from the specimen holder.

The low-power microscope designed by Musschenbroek was of a completely different pattern, but again there was provision for interchange of the lenses to vary the power. Each lens was mounted in a circular cell which could be attached to the end of an arm. This, shown

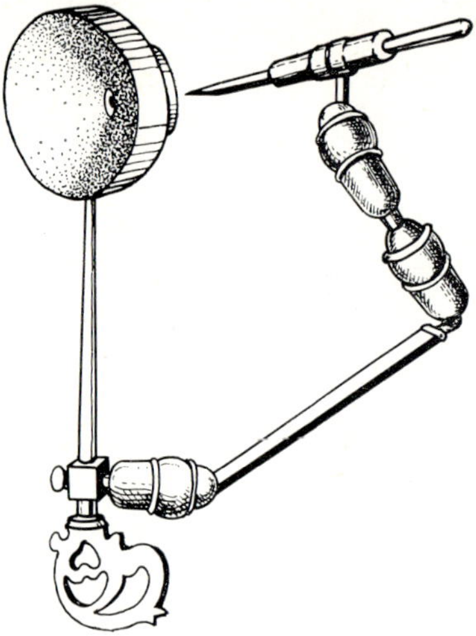

FIG. 3.6. Musschenbroek's low-power microscope. The lens is mounted at the base of the recessed eye-cup shown at the top left of the illustration. The object is impaled upon the pin which is carried on an arm. This in turn is adjustable by means of three ball and socket joints or "Musschenbroek nuts", so that the position and focus of the object may be changed.

in Fig. 3.6, served as a handle and hinged to this by a ball and socket joint was a second arm which carried the specimen support by means of further ball and socket joints. Several lenses were provided, six usually forming a set, with powers ranging from about 8× to 70×. Measurements of the optical performance of several of these lenses in the Leyden museum show that they are of good quality, the higher power ones especially.

Ball and socket joints to allow for the adjustment of the position of the specimen or the lens came to be widely used and the device was referred to as "Musschenbroek's nuts".

Lyonet, in particular, used them to carry the lens in the dissecting microscope which he designed around 1740 (Fig. 3.7).

The movement of the lens over the glass table was made easy by the use of Musschenbroek's nuts but this very mobility made focusing

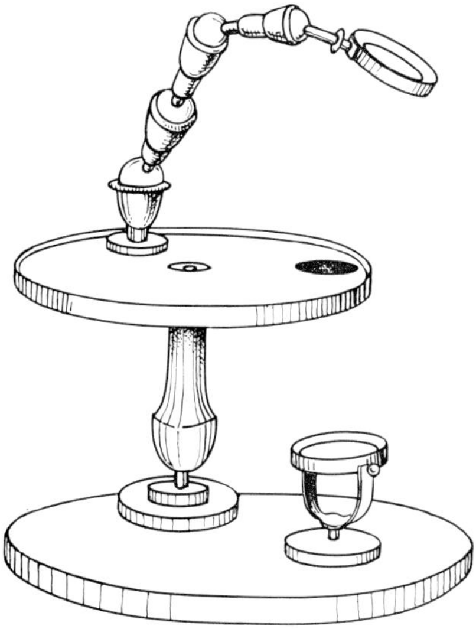

FIG. 3.7. Lyonet's dissecting microscope. The lens is carried on an arm composed of several "Musschenbroek nuts", so allowing its positioning with respect to the specimen.

difficult, especially with the higher power lenses. In order to get round this difficulty, Lyonet used to bend the arm carrying the lens a little too close to the object and then insert a tapering wooden wedge underneath the arm to elevate it very gradually until the point of correct focus was reached. At about the same time Musschenbroek nuts were used to provide a flexible mounting for accessory lenses used with compound microscopes and intended to serve for concentrating light upon the upper surface of an object. Often other accessories such as stage forceps were also provided with this type of flexible mount.

In the high-power design of Musschenbroek we have the focusing arrangement in which the lens holder and the object holder are pivoted and can be separated or brought nearer together to effect the focusing. This movement resembles that which takes place between the two legs of a pair of compasses and hence simple microscopes built on this plan have come to be known by historians of science as "compass microscopes". One form was sold by Wilson with the first model of his screw-barrel microscope; it was intended to accept the

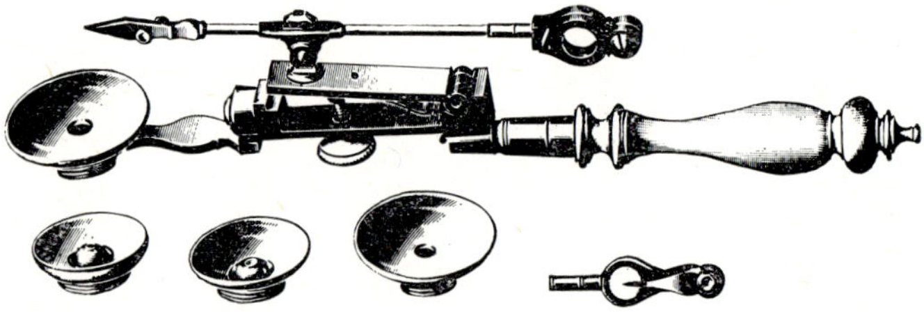

FIG. 3.8. A compass microscope probably made by George Adams about 1745. This instrument has a brass frame with an ivory handle and possesses interchangeable lenses, each of which is surrounded by a reflector or "Lieberkühn". The specimen was held in the forceps, or attached to the opaque stage (shown at the bottom right) which fitted onto the other end of the specimen rod.

same lenses as the screw barrel but to serve as an auxiliary microscope for the study of opaque objects. All the compass microscopes were intended to be held in the hand and by the middle of the eighteenth century they proved to be very popular instruments as they required very little skill to use.

Fig. 3.8 shows the typical form of a later type of compass microscope in which the lenses are inserted in the centre of a silvered reflector which acted as an illuminator for opaque objects; the analogy to a pair of compasses is clear from this illustration. One drawback of this type of instrument is that the object and the lens will move in an arc in relationship to each other as the focusing is carried out and hence their relative heights will vary. This was usually overcome by the device, already described in connection with the high power Musschenbroek microscope, by which the object-holder rod is made to slide within a sleeve and so can be brought into the correct viewing position. The addition of the silvered reflector for the illumination of opaque objects was made in 1738 by a German named Lieberkühn and such a means of illustration of an opaque object is still known by his name today. The device was not new, however, having been first described by Descartes in his *Dioptrique* in 1637 (see Chapter 1, p. 14) and by others since his time. The silvered reflector has the lens mounted in its centre and the focus and figure of the mirror are so adjusted that when any object is sharply focused by the lens, then it also lies at the focus of the mirror and so will appear brightly illuminated by the light reflected on to it.

The majority of the early compass microscopes had a simple friction joint between the legs, so that precise focusing was not easy. As, in general, the lenses were of low power this did not prove too great a disadvantage in practice. Wilson's first form of compass microscope (Fig. 3.9) had a crude form of screw thread focusing which must have been an improvement on the friction joint, but even this did not prove entirely satisfactory for it is criticized in the Philosophical Transactions of 1703 by an anonymous writer in letters addressed to "Sir C. H." (probably Christopher Holt). This writer is full of praise for the lenses of Wilson's microscopes but would have the screw thread replaced by one of much finer pitch in order to give a smoother focusing motion.

It is interesting to note, in passing, that the writer praises Wilson's lenses highly, rating them better than those of Mellen who, as we have seen, was much esteemed by Nehemiah Grew:

> As for the glasses themselves, I think them very good and well wrought and (tho' not so neatly set) go far beyond any I have seen of Mellins.

He did not, however, place these lenses of Wilson's in the same class as those of Leeuwenhoek:

> But the best of ours must needs fall short in power and goodness of Mr. Leeuwenhoek's Glasses, whose skill both in making and using them I fear we shall not easily reach.

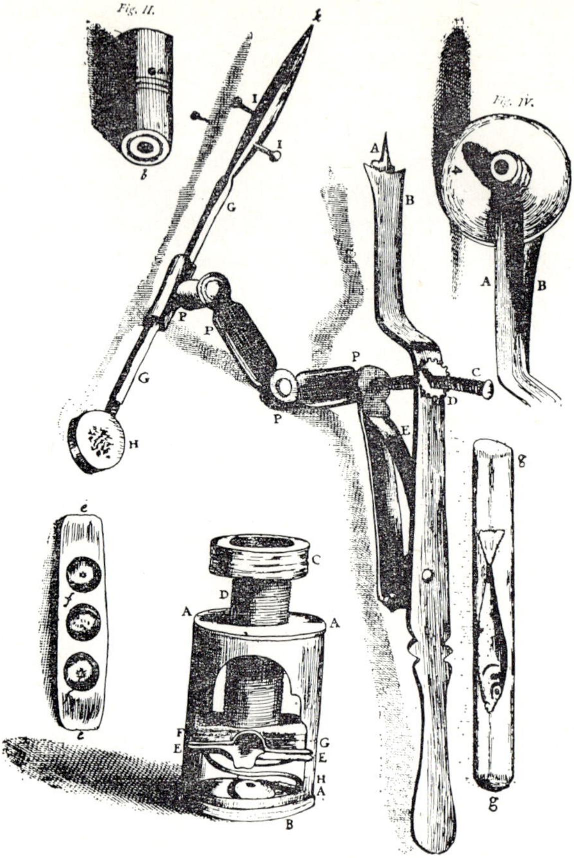

FIG. 3.9. Wilson's first model of the screw-barrel microscope of 1702. The screw-barrel microscope is shown in the centre of the illustration at the bottom. The focusing screw D is clearly shown, the specimen being placed between the plates EE and FG. The lens is mounted in the centre of the bottom plate labelled B. The compass attachment is shown at the right of the illustration with the arm AB, upon which were fixed the lens plates; the specimen rod and forceps are on the jointed arm at the left. In the upper right hand corner of the illustration one of the lenses is shown in position on the arm AB, whilst the lowest power lens is at the upper left. Also shown in the figure are an ivory slider with mounted objects and a tube containing a small fish ready for the examination of the circulation of the blood.

Later forms of compass microscopes such as that sold by John Cuff and the one made by George Adams, illustrated in Fig. 3.8, had a much improved form of screw focusing and were of a much higher standard of manufacture but still were not of much value for any serious study and they remained in use probably more for their simplicity in operation, serving to provide an easy way of observing some of the marvels of the minute world of Nature which were being revealed so rapidly in the seventeenth and early eighteenth centuries.

The introduction of the screw-barrel microscope at the beginning of the 18th century provided a cheap, portable microscope which above all was much simpler and easier to use than any of its predecessors; it had for the first time a reliable method of varying the power by means of interchangeable lenses and was provided with an accurate system of focusing the object. The original idea for the screw-barrel system of focusing was taken from earlier continental developments, involving compound microscopes. As far as can be determined the credit for the original idea must go to an Italian inventor named Tortona, who described microscopes embodying this system to the meeting of the Physics Mathematical Roman Academy held in 1685. Tortona, however, was reticent as to the methods of construction of these microscopes. The designs of Tortona were probably copied very closely by workers such as Campani and Bonanni, who have already been mentioned in Chapter 1. The latter figures (in his *Micrographia Curiosa* of 1691) a compound microscope with screw-barrel focusing to which his own spring stage had been added.

The application to a simple microscope seems to be due to Nicolaas Hartsoeker, about 1694. His drawing shows that in all essential features the microscope which gained such popularity in Wilson's name is that designed by Hartsoeker eight years previously. In fairness to Wilson, who has been accused of direct plagiarization, it should be stated that in his first article in the *Philosophical Transactions* for 1702 he makes no claim to be the inventor. James Wilson, who was in business "At the Willow Tree in Cross Street, Hatton Garden" was a typical practising optician of the time.

> He makes all sorts of Dioptric and Catoptric Glasses and Tellescopes, Prospects, Camera Obscuras, Magic Lanterns and Selleth the best of Spectacles and Reading Glasses.

His article in the *Philosophical Transactions* makes it clear from the title that he lays no claim to originality, reading,

> The description and manner of using a late invented set of small pocket-Microscopes, made by James Wilson; which with great ease are apply'd in viewing Opake, Transparent and liquid Objects; as the Farina of the Flowers of Plants, etc. The Circulation of the Blood in living creatures etc. The Animalcula in Semine, etc.

The screw-barrel type of instrument (Fig. 3.9) has the lens mounted in a circular disc which screws into one end of the barrel which forms the main body of the instrument. Within the body two thin brass plates are mounted so that they can slide up or down. Each plate has a central hole in it and they are pressed towards each other by the action of a strong spiral spring inserted into the body between one plate and the lens holder. The ivory sliders containing the objects to be studied or a small tube containing a living animal may be inserted between these two brass slides, when the pressure of the spring keeps the object firmly held. The two stage plates, as we may call them, together with the included object, are firmly pressed into contact with the "screwed barrel" (from which the name of this type of microscope is derived); this in turn is screwed into the main body of the instrument from the end opposite to the lens mounting. The amount to which this barrel is screwed in regulates the focus by pushing the stage plates and the object between them nearer to the lens, acting against the pressure of the strong spring. The basic features of this type of construction may be seen in Fig. 3.9 at the bottom centre.

In later years Wilson made several minor modifications to his design. In a pamphlet published in 1706 he mentions a turned brass handle which screwed onto a threaded fitting attached to the body of the screw-barrel microscope; it is clear that he altered some of the constructional details such as substituting leather for cork in the plate FG which pushed the brass stage plates together and by making the spiral spring of steel. Wilson's compass microscope was of better construction, and the lenses were now attached to it by screwing them into a ring. This pamphlet, together with its plate, were reproduced by Harris in the second volume of his *Lexicon Technicum* published in 1710 and furnished the basis for most descriptions of the Wilson microscope.

Some time between 1706 and 1710 Wilson published a further pamphlet which used the letterpress of the 1706 edition (and hence bears its date) but had a picture of his new opaque holder substituted on the plate. This rendered the compass attachment obsolete and hence it was no longer provided. The other minor point of difference was that the handle had a male screw thread instead of a female one and was now turned from ivory.

FIG. 3.10. A later model of Wilson's screw-barrel microscope, probably made by George Adams around 1746. At the extreme right of the picture is the lens holder for use when examining opaque objects; extra lenses and ivory sliders are shown in front of the actual microscope.

In this latter form, the screw-barrel microscope entered on its long career; such an instrument is seen in Fig. 3.10, with the handle attached to the screw barrel and an ivory slider in position. Extra lenses in their mounts, together with spare sliders are seen surrounding the instrument and on the right of the illustration is shown the opaque holder. This fitted to the barrel and the lenses were screwed into it, so that they were in effect outside the instrument. Culpeper about 1720 introduced the forceps plate which inserted between the stage and served to hold opaque objects for examination. A further innovation of Culpeper was the mounting of the body on a tripod stand to convert it from a hand-held instrument into a small table model.

There is no doubt that the development of the Wilson type of screw-barrel microscope contributed largely to the popularity of the simple microscope among the eighteenth-century amateurs. In addition to the superiority conferred on it by the very fact that it was a single lens microscope (and so did not suffer as much as the compound microscope from chromatic aberration), the screw-barrel microscope was of small size and therefore very portable. This presents a marked contrast to the microscopes of makers such as Marshall which have already been described in Chapter 2 and the Culpeper microscope which succeeded the latter. In addition, the screw-barrel microscope was very easy to use, almost the only precautions necessary in its use being succinctly summarized by Wilson himself:

> In the viewing of Objects, one ought to be careful not to hinder the Light from falling on Them, by the Hat, Perruke, or any other thing, especially when they are to look upon Opake Objects: for nothing can be seen with the best of Glasses, unless the Object be in a due distance, with a sufficient Light.

Harris in the preface to the second volume of the *Lexicon Technicum* praises Wilson's microscope very highly:

> of all the Microscopes I have ever seen for Commodiousness, various Uses, Portability, and Cheapness, I never met with anything like Mr. Wilson's glasses.

This opinion was evidently endorsed by the public and the screw-barrel microscope was produced by all the instrument makers of the eighteenth century, and often at a very low price. A catalogue of Henry Pyefinch, dated 1775, shows that although he was producing his best compound microscope with a jointed pillar for 7 guineas, a Wilson screw-barrel could be bought for only 2 guineas, or £2. 10s. 0d. if the adaptor for opaque objects was included.

At this time the single-lens microscope was proving of great value in the study of the fresh-water coelenterate animal *Hydra* first described by Leeuwenhoek. During the eighteenth century this creature was usually referred to as the fresh water polype, and in 1739 Abraham Trembley published some account of his discoveries relating to this animal. This was followed in 1744 by his famous book *Memoires pour servir à l'histoire d'un Genre de Polypes d'eau douce, à bras en formes des cornes* in which he pointed out the remarkable regenerative power of *Hydra* and proved experimentally that they could produce a complete individual from either portion of an animal which had been cut in two. Working with the simplest of equipment (a glass jar full of water and weed containing the polyps which were examined by a simple lens mounted on an arm of Musschenbroek nuts so that it could be moved to examine any part of the front surface of the glass jar) Trembley also discovered their methods of feeding and reproduction. Such work stimulated others, particularly Henry Baker in this country, to observe these animals.

Baker produced *An Attempt towards a Natural History of the Polype* in 1743 and in 1755 Ellis published his essay on the natural history of the corallines which was to become justly famous. This particular type of microscopical work imposed certain demands upon the instrument and led to certain innovations. As the slightest jar or vibration could cause the *Hydra* to contract and withdraw its tentacles, the tank in which they were kept had to remain quite still. Hence the lens had to be moved over the tank rather than the specimen under a fixed lens as had been the custom.

Again, as the tanks were often of considerable size, the lens traverse had to be of a considerable extent. Such requirements led to the development of a microscope which had a large stage on which the tank could be placed and a lens attached to an arm. The arm was fixed at right angle to the pillar of the microscope and could be moved not only to bring the lens nearer to or further away from the optical axis of the microscope, but also to pivot and so enable the lens to be traversed over any part of the tank. Such microscopes with this type of movement for the lens became known as "aquatic" microscopes (Fig. 3.11) in distinction to those in which the lens could only move up and down with respect to the stage. From their chief use these latter were generally termed "botanical" movements.

The original form of Ellis's microscope was probably constructed by Cuff; Ellis tells us:

> I have used a very commodious microscope of Mr. Cuff's, the Optician in Fleet Street, which I had altered for that purpose.

This instrument is shown in Fig. 3.11, from which the general construction may be seen. The large stage, focusing method, and device for obtaining the aquatic movement and the large mirror are all obvious.

The method of mounting the pillar in a bush screwed onto the lid of the box was a very common adaptation for portability and occurred frequently in instruments built at this period and for the next century or so. Ellis's aquatic microscope may be regarded as the direct ancestor

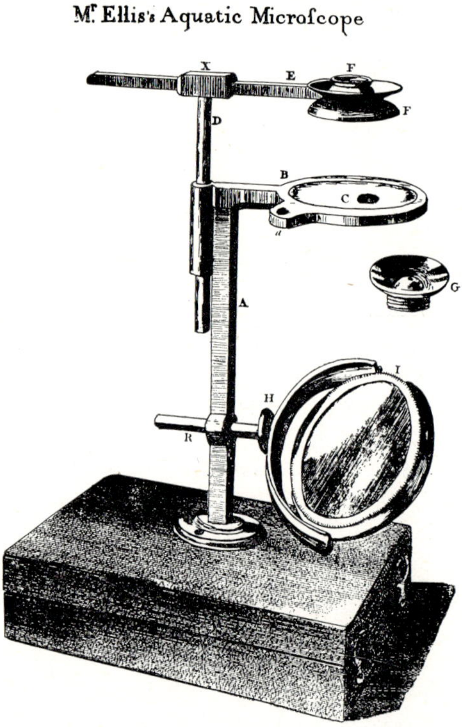

FIG. 3.11. "Mr. Ellis's aquatic microscope". The aquatic motion was provided by sliding the arm E in the socket X, and by swivelling the rod D in the mount attached to the main pillar labelled A. The stage (C) and the mirror are also shown. The lenses, one of which is drawn separately, are provided with Lieberkühn reflectors.

of the standard low-power dissecting monocular in use today in schools; this has now the advantage, however, of an achromatic doublet or triplet magnifier. Such an instrument (Fig. 3.12) provides a very adequate means of obtaining magnification of a few diameters for dissection and biological preparation work in general.

The majority of the eighteenth-century users of the simple microscope, like Ellis, were content to use rather low-power lenses for their

FIG. 3.12. A modern simple dissecting microscope, with rackwork focusing and glass stage plate.

studies, although there were notable exceptions, such as the botanist Robert Brown in the first half of the nineteenth century. In general, however, following the death of Leeuwenhoek there was a lapse in interest in single lenses of high power. This was probably due to difficulties of construction of such lenses and to the discomfort of working with a magnifier which requires a close working distance and has a very small diameter.

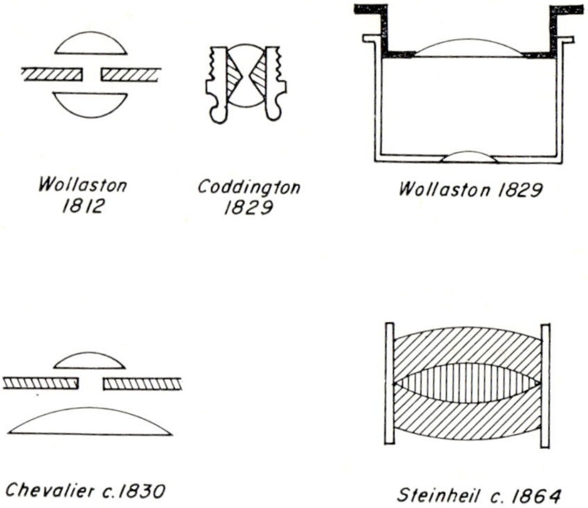

FIG. 3.13. Diagrams of the lens arrangement in various magnifiers of the nineteenth century.

Distrust of the images produced by the compound microscope persisted throughout the eighteenth and the early years of the nineteenth century, so efforts were devoted to improving the single-lens microscope.

Wollaston, in 1812, devised a system whereby two plano-convex lenses were fixed with their plane surfaces together, separated by a small diaphragm. This arrangement, indicated in Fig. 3.13, Wollaston considered would largely mitigate the spherical aberration present in high-power single lenses. Other systems, such as the Stanhope lens and the Coddington lens followed in succeeding years (see Fig. 3.13), but none of them could be considered entirely successful. In order to reduce the spherical aberration to an acceptable figure the diaphragm had to reduce the working aperture of the lens system very considerably.

About this time, however, rapid strides were being made in the development of achromatic glass lenses, which resulted in a revival of interest in the compound microscope as a tool of scientific research.

The single-lens microscope was then to some extent relegated to the role of a low-power naturalist's microscope and to that of a dissecting instrument which served to prepare material for study with the compound microscope. Even though the new achromatic lenses which were coming into fashion in the late 1820's had a better resolving power than single lenses of equivalent power, some workers carried on using the old simple microscope for some time. Coddington in 1829 had developed a form of magnifier, known as the "Bird's-eye object glass" in which the use of an opaque diaphragm, as shown in Fig. 3.13, reduces the aberrations; about the same time Wollaston produced a fresh design for a doublet which mitigated still further the inconveniences of small field and great degree of spherical aberration which beset the simple lens. This design was taken up and further improved by Chevalier and still later by Zeiss who, in the early part of his career, specialized in this type of lens. In Chevalier's doublet the use of a diaphragm between the lenses (Fig. 3.13) sharpened up the image and later the lower lens of Wollaston's doublet was replaced by a combination which enabled the aberrations to be still further corrected. Van Cittert, who has measured several Wollaston doublets, quotes one made in 1835 by Dollond of London which had a magnifying power of 200 × and a resolution of about 6 μ and another with a power of 300 × and a resolution of 5 μ.

In the 1830's these achromatic triplets were far superior to the early simple microscope and they set a standard which has been continued and improved upon to the present day. Achromatic lenses were being applied to the compound microscope, but good work was still being produced by single-lens microscopes; von Baer in 1827 described and discovered the mammalian ovum, and Robert Brown carried out his remarkable botanical studies with such an instrument. He is better remembered today as the discoverer of the movement of small particles in a colloidal solution as the result of molecular bombardment, but he studied many families of plants in great detail. In 1833 he published

"Observations on the organs and mode of Fecundation in Orchidaceae and Asclepidaceae" in the *Transactions of the Linnean Society*, a paper which, despite its name, contains observations of profound significance. With the aid of a simple microscope he had been able to see and describe what we know as the nucleus of the cells. Although the cell concept had not been clearly enunciated in 1833, Brown was able to recognize that here he was dealing with a structure of great importance in the life of the plant, a structure which proved so important that it is still a vital object of study to microscopists in the middle of the twentieth century.

Although high magnifications can be obtained from a single lens, the resolution and light gathering power of such lenses are far surpassed by those of a modern objective of similar magnification.

At the present time the use of the simple microscope is largely confined to very low-power work, magnifications ranging from 5–20 diameters. For this purpose the combination of lenses invented by Steinheil in 1864 is largely used. By the use of glasses of different dispersions and differing curvatures it has proved possible to produce an image which is aplanatic, achromatic and free from astigmatism. In addition the whole combination has a reasonable working distance and a large field of view. For dissection work involving powers of from 20–100×, the compound binocular microscope built according to the design of Greenough, i.e. having two completely separate microscope systems, one for each eye, is now largely used.

The developments of the microscope during the eighteenth century took place mainly in England, and were largely concerned with the mechanical part of the microscope. During this time England was accepted as the centre of microscope progress, a role which she maintained until the latter part of the nineteenth century, when as a result of the work of Ernst Abbe, Germany took the lead in the design of microscope lenses.

CHAPTER 4

The Eighteenth Century—Mechanical Progress and the Achromatic Microscope

THERE can be little doubt that the writings and researches of the pioneer microscopists, such as Hooke, Grew and Leeuwenhoek, stimulated a tremendous interest in the minute world which lay all around. The new tool caught the imagination not only of the natural philosophers, but also of the layman; towards the beginning of the eighteenth century it was a status symbol to possess a microscope and be able to demonstrate with it the structure of insects, flowers and other natural objects. This development was of profound significance for the future of the microscope, for it resulted in the development of new types of instrument.

The professional scientists, however, tended to view the results of investigations carried out with the microscope with some scepticism, probably partly justified; the poor optical corrections of the lenses and the very low apertures then in use resulted in images which were indistinct and often surrounded by colour fringes. This in its turn probably caused many structures to be described which were in fact false, resulting solely as optical effects produced by the instrument. A good example of this may be found in the microscopical study of the spermatozoa which throughout the eighteenth century were often pictured as containing a minute "homunculus" or fully-formed miniature of the human form. This faulty interpretation was very often due also to the mistaken ideas of the observers. Such errors of interpretation and faulty specimen preparation, together with those due to the instrument itself, were responsible for the view that the microscope was only of value for the lowest power work; in the eighteenth and early nineteenth centuries the term "microscopical deception" was much used to signify the suspicion and mistrust which was often felt for results of microscopical research.

The amateur tradition, however, became well established in England and from this in the nineteenth century was to spring a tremendous fount of inspiration and enthusiasm which led to notable technical advances in the optics of the microscope. These amateur users of the instrument, by their very great skill (often far exceeding that of the professional scientist) in using the microscope at its limits of resolution contributed very greatly to the development of the modern instrument and to our present techniques.

The large popularity of the microscope in the eighteenth century was to a great extent catered for by the Wilson screw-barrel instrument, which was cheap and easy to use. Many people, however, wanted a much larger microscope. At first this would undoubtedly have been a Marshall type, as in the case of Samuel Pepys (quoted in Chapter 2). This was large and cumbersome but very versatile in skilled hands. Between the years 1725 to 1730 a new microscope appeared on the market, of a type which was to sweep away the larger, older model and which was to persist in one form or another for the next hundred years.

This new microscope was designed by Edmund Culpeper and it departed from the single pillar type of construction favoured by Hooke and Marshall in favour of a modified form of the tripod used earlier by Campani in Italy. This basic similarity is obvious if Campani's instrument (Fig. 1.14) is compared with a typical Culpeper microscope shown in Fig. 4.1.

In its basic form, the Culpeper microscope consists of a round base or stand of wood, from which three turned legs rise to support the stage. From this in turn three further legs arise to terminate in a brass ring supporting a sleeve. In early models this tube or sleeve was constructed in the typical fashion of the period, i.e. a cardboard tube covered with leather or ray-skin. This tube acts as a support in which the actual microscope body slides in order to provide a focusing movement. Later a draw tube was often provided. From the illustration (Fig. 4.1) the general appearance may be gauged. The most obvious feature is the eyepiece portion of turned wood which bears a very strong resemblance to those of the Marshall type instruments. The eyepiece would have contained a bi-convex eye lens, together with a larger field lens situated at the lower end of the turned wood portion. Another very striking feature is the disposition of the legs supporting the sleeve which are arranged to come at intermediate positions between the lower legs supporting the stage itself. Early forms of the Culpeper

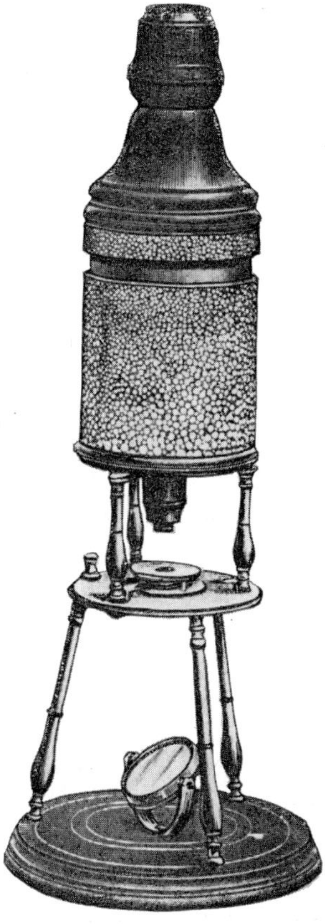

FIG. 4.1. A typical Culpeper microscope. Note the relative positions of the tripod legs supporting the stage and the body tube. The instrument is fitted with a Bonanni spring-stage.

microscope usually had a box cap, designed to hold the spare objectives, which screwed onto the eyepiece and served as a dust excluder.

One very important innovation can be seen from Fig. 4.1; in the centre of the wooden base is fixed a concave mirror in gimbal mountings so arranged that it can be set to reflect light up through any object placed upon the stage of the microscope. It seems probable that earlier workers had in fact used a mirror; the inconvenience of the Campani microscope in which the whole tripod had to be held up to the light, or that of the Marshall where the stage and the body had to be swung over the edge of the table in order to bring the optical axis over a candle placed on the floor, would certainly have resulted in some attempt to use the principle of reflection from a mirror to alter the direction of the illumination.

Any such experiment, however, would have used a plane mirror. Culpeper's idea was to use a concave mirror which not only acts as a reflector but also concentrates the light and so acts to some extent as a condenser. This eliminated the extra bi-convex lens which was used for this purpose in the Marshall instruments. The placing of the mirror in the centre of the base of the Culpeper microscope directly underneath the optical axis was a most important development and this probably resulted in the perpetuation of the upright, non-inclinable form of microscope for many years. In the sheet of instructions which Culpeper supplied with his microscope he made the specific point:

> remember the Concave Looking-Glass at the Bottom is to reflect the Light up to the Object either by Day or Candle-Light (a bright white Cloud gives the best Reflection by Day)

This seems to suggest that the concave mirror was in fact an innovation, otherwise it would not have been necessary to stress in detail what very soon was accepted as an obvious function.

Like the earlier Marshall microscopes, the Culpeper instruments were provided with different powers, usually five in number, each mounted in a cell or "button" made to screw onto the small tube at the lower end of the main microscope body. Van Cittert has measured the optical performance of the instrument when fitted with each of these objectives in turn and he found that the magnifying power varied from about 30× with the objective number five, 60× with number four, through 80×, 100× to 275× with the highest power. All these lenses, of course, were uncorrected for chromatic aberration and were provided with diaphragms behind the lens which had only a small opening in order to minimize spherical aberration. In consequence, the aperture was low and the resolution was poor. Measurements with test plates suggest that the highest power lens had a resolution of the order of 5 microns (1 micron or $\mu = 1/1000$ mm) whereas a modern objective lens combination which would be used to produce the same magnification would be expected to resolve about $0 \cdot 5\ \mu$.

The Culpeper form of microscope stand would undoubtedly have been easy to construct, and this would have helped towards the production of a comparatively cheap instrument. At first the stage was constructed of wood, but very soon Culpeper turned to the use of brass and all his later models have their stages made of this metal. The final modification of this item was the addition of a type of bayonet joint to the aperture in the centre of the stage so that a Bonanni spring stage could be added to hold the prepared ivory sliders firmly. Most of the instruments were provided with a fish plate, often made entirely of glass as in the earlier microscopes, in which case the fish plate was often elaborately engraved with the purpose of the accessory,

> this Glass is to lay ye Fish on.

In the latter models the fish plate was of glass edged with brass or again constructed entirely of metal. In this case the plate had a curved shape, something resembling a shoehorn, with a slot cut at one end over which the tail of the fish was to be laid.

Although the Culpeper microscope was easy to make, was relatively cheap and had the advanced concave mirror design, it nevertheless represented a definite regression in microscope design. This is particularly true when one considers some features of detail, such as the fact that the access to the stage was severely limited by the arrangement of the tripod legs supporting the microscope body. This limited the use of the instrument almost entirely to the study of objects by transmitted light, although the maker was reluctant to admit this and even provided a place for mounting a bi-convex lens on the edge of the stage for illuminating opaque specimens. Again it must be noted that the inclining limb of the Hooke and Marshall microscopes has disappeared, so that the Culpeper microscope can only be used in the upright position. Finally, the fine focusing adjustment to be found on the Marshall instruments has been abandoned and the instrument is focused solely by sliding the draw tube in the sleeve.

Instruments of the same basic pattern were made by the majority of the instrument makers of the period; particularly fine examples were produced by Matthew Loft who was active between 1720–1747. His microscopes are notable for their fine workmanship and for their decoration. It is typical of the period, when the beauties of design in furniture and the other arts were really appreciated, that the microscope should also receive its share of decoration and become an object of beauty as well as use. Certainly the bodies covered with the prepared ray-skin and the polished brass and mahogany of microscopes of the Culpeper pattern result in an extremely attractive instrument, although one which was not very practical to use. As the tubes were made of cardboard, the fit of one within another would be constantly changing, and to a large extent dependent upon the weather conditions; the only real solution to this problem was to construct the whole instrument of brass, which was eventually achieved by John Cuff.

Culpeper microscopes with brass bodies and rackwork focusing were produced for over a hundred years. Illustrations of this type appear regularly in almost every book on microscopes published in this time and Adams in fact refers to this form of instrument as the "Common Three-pillared Microscope". Peter Dollond was making this pattern of microscope in 1780 and it is likely that examples were in production until well into the nineteenth century.

Although there were many mechanical improvements in the design of the microscope in the early part of the eighteenth century, it is a pity that the basic principles of microscope design which involved the mounting of the body and the stage on an inclinable pillar, so ensuring their constant alignment, were abandoned for the inconvenient tripod form. This latter form persisted as a negative influence on instrument design for far too long, and the real break-through did not come until the emergence of Benjamin Martin and George Adams the Elder as the two great figures in microscope design and manufacture in the second half of the eighteenth century. Before we turn to consider their contributions to the evolution of the microscope, the instrument which

John Cuff himself designed must be mentioned. It was probably due to the instigation of Henry Baker (1698–1774), a most interesting figure in the scientific life of the period.

Baker carried out experiments on the crystallization of salts under the microscope, for which work he was awarded the Copley medal of the Royal Society in 1744; at first he used a Culpeper type of microscope but he found that the legs were a great nuisance for this type of experiment. In his book *Employment of the Microscope* (1753) Baker tells us that

> The cumbersome and inconvenient Double Microscope of Dr. Hook and Mr. Marshall were many Years ago reduced to a manageable Size, improved in their Structure, supplied with an easy Way of enlightening Objects by a Speculum underneath, and in many other Respects rendered agreeable to the Curious, by Mr. Culpepper, and Mr. Scarlett. Some farther Alterations were however wanted to make this Instrument of more general Use, as I fully experienced in the Year 1743, when examining daily the Configurations of Saline Substances, the Legs were continual Impediments to my turning about the Slips of Glass; and indeed I had found them frequently so on other Occasions. Pulling the Body of the Instrument up and down was likewise subject to Jerks which caused a Difficulty in fixing it exactly at the Focus: there was also no good Contrivance for viewing opake Objects. Complaining of these Inconveniences, Mr. Cuff, the Optician, applyed his Thoughts to fashion a Microscope in another Manner, leaving the stage intirely free and open by taking away the Legs, applying a fine threaded Screw to regulate and adjust its Motions, and adding a concave Speculum for Objects that are opake.

Baker goes on to give a drawing and a description of the resulting instrument which he refers to as "Mr. Cuff's new-constructed Double Microscope". From a pamphlet by Cuff, dated 1744, we learn that although he had attempted to achieve his purpose by modification of the traditional Culpeper stand he was not able to claim success in this way. Cuff continues:

> But the legs I could not get rid of, nor was it possible in that form to make the whole fully answer the desires of the curious. In order, therefore, to remove all complaints the present microscope is contrived on a new construction whose motion is easy and regular and steady; whose application for opaque and all other objects will be found convenient; whose stage is quite free for any object to be applied on; and whose general

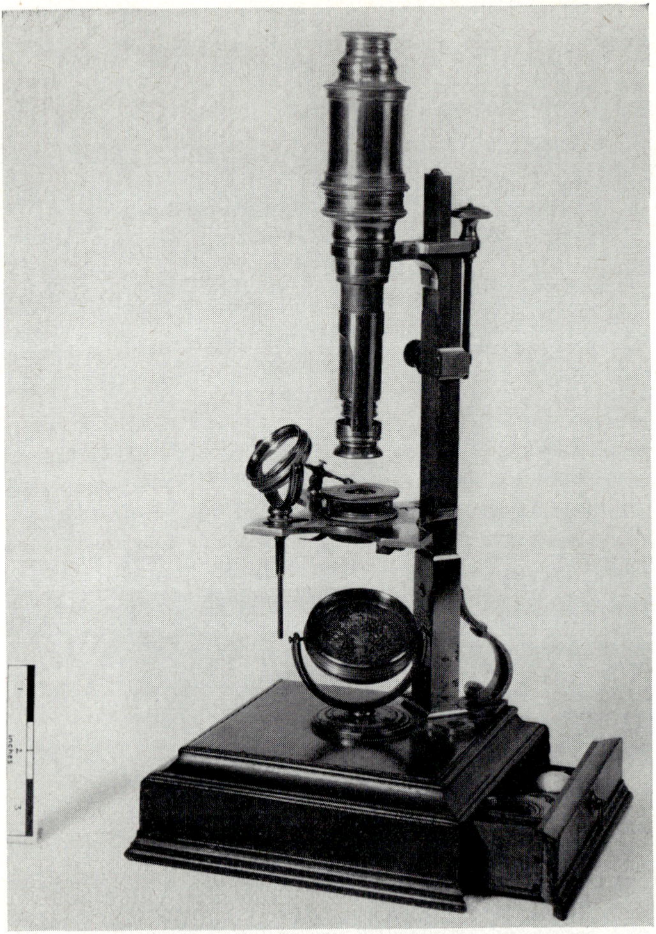

Fig. 4.2. A Cuff-type microscope in the Wellcome collection. This instrument is fitted with a Lieberkühn reflector mounted at the end of a tube, which slides upon the parallel nosepiece, and with a condensing lens providing incident illumination. The fine-focus screw is visible at the rear of the pillar of the instrument.

(Copyright, The Wellcome Trust and by courtesy of the Wellcome Trustees.)

figure cannot, it is hoped, be thought unhandsome; but this is all humbly submitted to the opinion of the best judges on whose favour I rely and whose obedient and most devoted servant I am, John Cuff, Fleet Street, 20th September, 1744.

There is no doubt that this new microscope of John Cuff represents a decided step forwards in the evolution of the microscope stand, although of course optically it was no different from what had gone before. The general form of the instrument may be seen in Fig. 4.2 and some of the detailed construction in Fig. 4.3.

Cuff's microscope was a sound design and it achieved a great success. Most of the instrument makers of that time paid it the compliment of using it as a model for their own copies, and it was produced in essentially the same form up to about 1800. The Cuff microscope had a neat appearance, it had a reasonable degree of rigidity and the focusing mechanism and access to the stage was good. It may well be that the immediate and sweeping success of the Cuff microscope was responsible for the rapid change in the design of the microscopes offered for sale by some of his contemporaries, especially George Adams the Elder.

One of the trends initiated about this period and which continued well into the nineteenth century was the portable microscope, which could be packed away and carried around very easily. Many special models were designed for this purpose but most of the current instruments were adapted as portables by some maker or other. Nairne achieved this with his version of the Cuff microscope by mounting the pillars on a hinge fixed to one end of the box so that they could swing down, taking the microscope stage and body with them into the box (see Fig. 4.4). This particular development of the Cuff model is often referred to as a "chest" microscope, and it not only has the virtue of making the microscope easily portable, but also allows more convenience in use by permitting the angle of the limb to be altered so as to bring the body and eyepiece to a convenient height for the observer.

In the fourth edition of the *Micrographia Illustrata* published in 1771, Adams includes a catalogue at the back. Here the "double constructed microscope", i.e. one of the Cuff type, is listed at £6. 6s. 0d., or £8. 8s. 0d. if supplied with a triangular foot. At the same time Adams

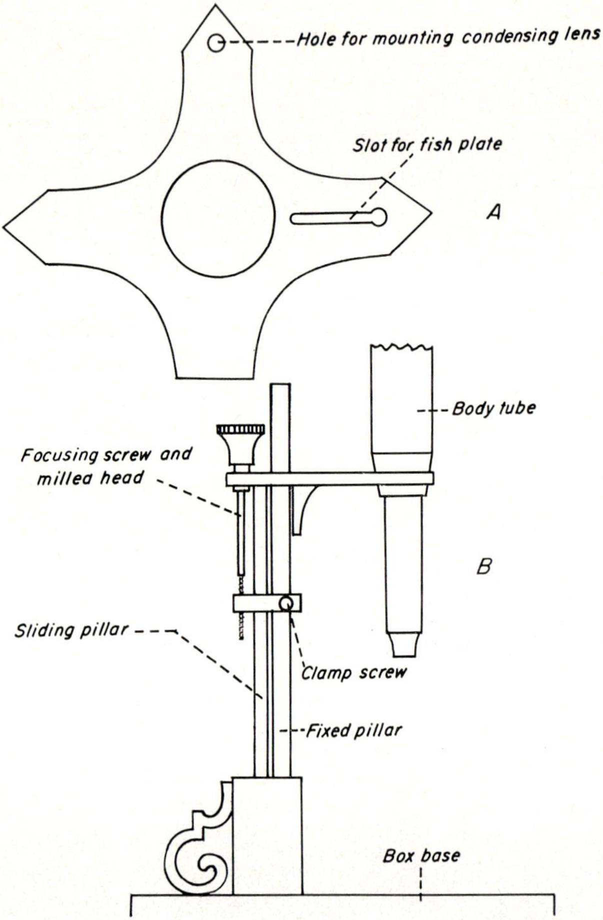

FIG. 4.3. A. The typical stage which was fitted to the Cuff-type of microscope.

B. Diagram to show the mechanical focusing arrangements of the Cuff microscope.

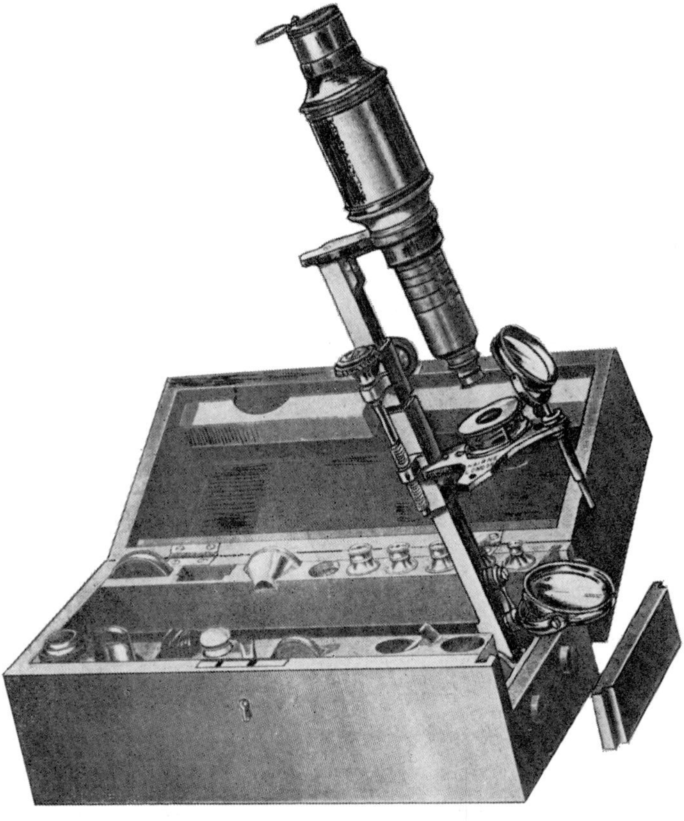

FIG. 4.4. A Nairne "chest" microscope. The pillar is hinged so that it will fold into the box to allow the instrument to be packed away easily for transportation.

was still supplying Culpeper microscopes at 3 guineas each, together with Wilson screw-barrel and Ellis's aquatic microscope both costing 2 guineas.

The development of the microscope throughout the remainder of the eighteenth century revolves very largely around the inventions of Benjamin Martin and the two Adams, father and son. George Adams senior was an optician and instrument maker by trade; his early life remains shrouded in obscurity but we know from one of his advertisements that he was established in business as a mathematical instrument maker at the sign of "Tycho Brahe's Head" in Fleet Street by 1735. He continued constructing instruments until his death in 1773 when the business passed to his elder son, also called George, who continued it in his turn. George Adams the Younger died in 1795, upon which his brother Dudley assumed the direction of the firm.

Benjamin Martin was almost exactly contemporary with George Adams the Elder, but started off in a very different manner. He was certainly well educated and probably had a formal mathematical and scientific training. Early in his life he was a schoolmaster and wrote extensively on mathematical, philosophical and astronomical subjects. In a book of his devoted to trigonometry he dates the preface "From my school in Chichester, April 8th 1734". Later in the same work he advertises himself as "Teacher in writing in all the common and useful hands, arithmetic, etc., etc., down to the use and construction of the most useful mathematical instruments". This suggests that he started making instruments and microscopes as a hobby, but by 1736 he had begun to offer them for sale to the public. From the catalogue and price list included in his *Description of the Globes*, we find that at this time Martin was selling a "large parlour compound microscope" for £3. 12s. 6d. and the same in brass for £5. 5s. 0d. A Wilson microscope and apparatus cost £2. 12s. 6d. as did "Dr. Lieberkühn's opake microscope".

Benjamin Martin proved to be a very fine craftsman and a skilled user of the microscopes which he made; he was, moreover, an original thinker. All these attributes contributed to establishing him as one of the notable figures in the development of the microscope during the eighteenth century.

One of his early instruments was the pocket microscope of 1738; this microscope was almost certainly made, as its predecessors had been, of cardboard tubes and wooden mounts. No example of this instrument is known but from the description we gather that the instrument would have been about six inches long when the draw tube was closed, with a diameter of about one and a half inches. The general form of the instrument is shown in Fig. 4.5, which is redrawn from Martin's pamphlet. The concave mirror which reflected light onto the specimen could be opened out for use and one further feature of interest is that the eyepiece was provided with a slider to act as a dust excluder, just as in the majority of telescope eyepieces of the time. This particular type of instrument may be regarded as the ancestor of the "drum microscope" which proved so popular and which was made by Martin for many years.

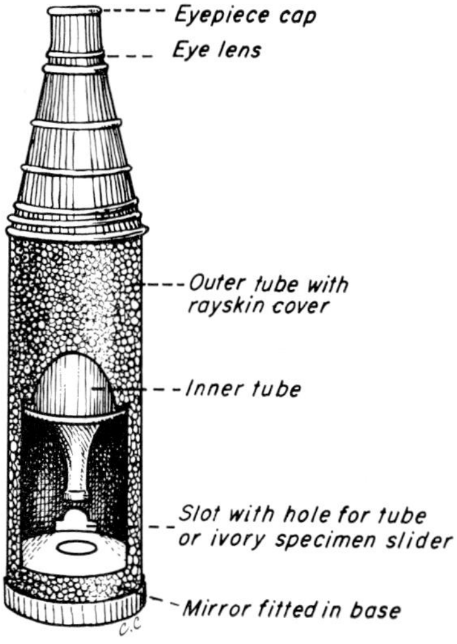

FIG. 4.5. Benjamin Martin's pocket microscope of 1738. This instrument was the forerunner of the famous drum microscopes.

Martin provided two objectives with the pocket microscope, one of focal length which he stated as $\frac{37}{100}$ inch and the other $\frac{66}{100}$ inch. The eyepiece lacked a field lens, consisting solely of a single biconvex lens. Martin obviously believed that the great merit of this microscope was its portability, and he describes in some detail how it may be adapted to screw into the hollow handle of a walking stick or rigid whip. The price of this new instrument was stated to be 12s. 6d. in its simple form or 1 guinea if a micrometer was fitted.

Although Martin gave very precise directions for using his microscopes, it seems that many people were still uncertain how to proceed and so only succeeded in obtaining indifferent results which they blamed not upon their lack of skill, but upon the faults of the microscope. In an appendix to his pamphlet Martin comments rather sourly:

> I could never have thought that the nature and use of the microscope (so exceeding curious and useful an instrument) had been so little understood at this time of day, in a land of so much Knowledge as England, that any difficulties could arise concerning the use of the Pocket Microscope, whose nature, construction and use are and must be allowed the most obvious, simple and easy of all others yet invented.

He then proceeds to explain the method of use yet again and stresses the importance of paying attention to the suitability of the illumination, adding:

> A very cloudy, dark air, or gloomy room, will always defeat your purpose, and make you have an ill opinion of the instrument without cause.

Shortly after these early instruments of Benjamin Martin had appeared, George Adams announced in his *Micrographia Illustrata* of 1746 the "New Universal Single Microscope" and the "New Universal Double Microscope".

These two instruments were designed to fill all the possible roles of a microscope; they could be used as a simple microscope, as a compound instrument or even be adapted to serve as a solar microscope in which the image is projected onto a screen (see p. 113).

The "Universal Single" microscope is illustrated in Fig. 4.6; it was made, as Adams tells us, "either of Brass or Silver, and is composed of six double Convex Lens's of different Foci". These lenses were mounted on arms which projected from a disc. This disc was capable of rotation around the axis of the pillar on which it was mounted, so that each lens in turn could be brought under the eye shield (labelled N in the illustration, which is taken from Adams's book). One of the lenses carried a Lieberkühn or as it was then termed "a reflecting Speculum of Silver or other Metal, highly polished, which when an Opake Object is to be viewed must be placed under the eyepiece N."

Focusing of the object is effected by moving the stage up or down the pillar by means of the action of a threaded rod which passes up inside the pillar and is actuated by turning the nut P at the base, at the

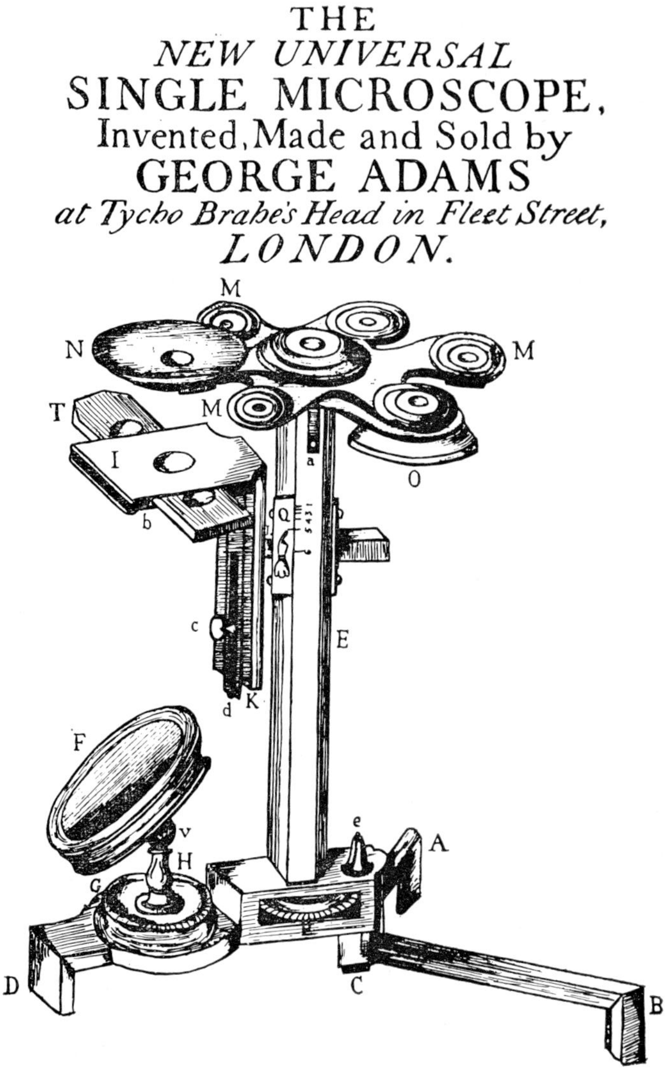

FIG. 4.6. George Adams's "New Universal Single" microscope of 1746. The fine focus control is at the base of the pillar and the wheel of lenses can be clearly seen.

centre of the tripod foot. Adams was pleased with the convenience of the focusing screw:

> I must here observe that the Screw P is to be turned as your Hands and Arms are resting upon the Table, which is a Convenience to be met in no other Microscope. All others require the Observer to raise his Body and Arms in adjusting the Object to fit his sight; which is not only very troublesome but tiresome too; especially if it requires considerable Attention. Whereas in this new Universal Microscope a leaning Posture is sufficient, and consequently the easiest of all others for Microscopical Observation.

This device is also found in the "Universal Double" microscope (Fig. 4.7) and although it did not become standard practice in the eighteenth or nineteenth centuries, it has reappeared as a feature of most research microscopes which are manufactured today, so that the arms may rest upon the table during operation of the instrument.

The Single microscope suffers from the drawback that owing to the diameter of the disc-shaped lens carrier (which is about four inches) it is rather difficult to place one's eye close to the eye shield without an awkward bending of the neck. Another fault in the design is its lightness; such flimsy construction makes it impossible to operate the focusing screw unless the microscope stand is firmly held down onto the table with one hand.

The "Universal Double Microscope" is of very similar construction. That shown in the *Micrographia Illustrata* carries the compound body in a ring attached to a short arm which slides up and down the pillar with the ring of lenses. The coarse focus was obtained by sliding this complex up and down, the rough position being indicated by numbers as in the single microscope. On clamping the socket the fine focusing mechanism was provided by an internal screw which acted upon the stage, in exactly the same manner as in the other instrument. This model of the Double Microscope must have been abandoned very quickly, for the examples which exist in our collections today represent a slightly modified version (Fig. 4.7) in which the compound body screws into the socket in place of the eye shield and the separate supporting arm is no longer provided. The Lieberkühn is now mounted on an independent slider working on the back face of the block which supports the stage. This allows focusing of the Lieberkühn quite independently of the lens.

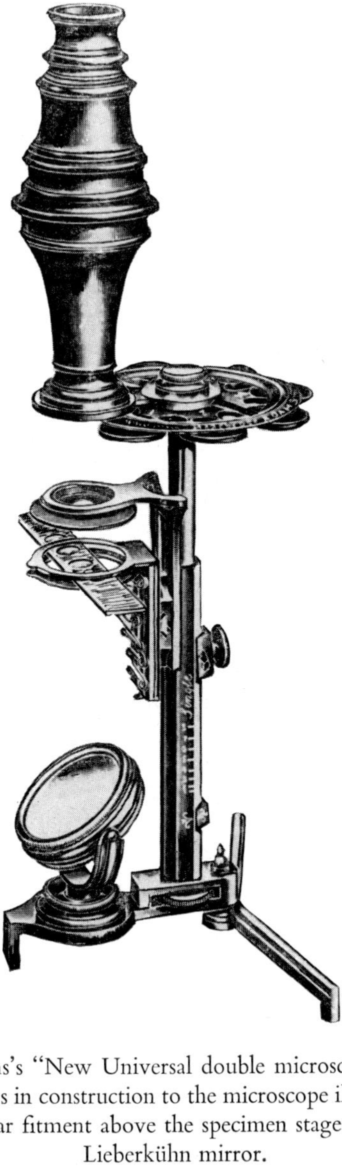

Fɪɢ. 4.7. Adams's "New Universal double microscope". Note the similarity of this in construction to the microscope illustrated in Fig. 4.6. The circular fitment above the specimen stage is an adjustable Lieberkühn mirror.

Both of these instruments reintroduced the sector folding foot, which Culpeper had used for his table stand for the Wilson microscope, but in these new microscopes of Adams the sector foot forms an integral part of the whole assembly and not just an optional extra. The main drawback of these microscopes must undoubtedly have been their lack of solidity.

Probably one of the best ways of increasing the rigidity of the microscope stand, whilst still retaining the flexibility and ease of operation conferred by the ability to incline the limb, is by supporting the whole limb at its centre of gravity between trunnions. This type of construction was in fact adopted by both Adams and on the continent by a famous French instrument maker named Magny. It does not seem possible to say which of the two makers had the priority, or whether they both independently came to the same decision, but in 1754 Magny was constructing stands of this pattern. He arranged for the whole limb to swing between trunnions, not with a fore and aft inclination as one might expect but laterally. Adams, on the other hand, when he adopted this type of construction for the microscope which he made for King George III whilst he was still Prince of Wales, used the more conventional idea of having the pivots on either side of the stand so that the inclination resulted in the body being brought nearer to the observer who is normally sitting behind the microscope.

Adams's "Prince of Wales" microscope is now in the Science Museum at South Kensington and it may be dated between the years 1751 and 1760, when the Prince of Wales ascended the throne. The stand, which is shown in Fig. 4.8, has a Cuff-type focusing mechanism with a wheel of objectives which vary from $\frac{1}{20}$ inch to $\frac{3}{4}$ inch focus, much as in the second version of the "New Universal Double Microscope".

This particular microscope may well be regarded as the peak of George Adams's achievements, certainly as far as a practical instrument is concerned, although for sheer craftsmanship and beauty the later microscope in silver (shown in Fig. 4.10) ranks very highly. As will be seen this latter instrument cannot be regarded as a useful scientific instrument, intended for serious observational work.

In the meantime Benjamin Martin was continuing his highly successful activities as an instrument maker. He had moved from

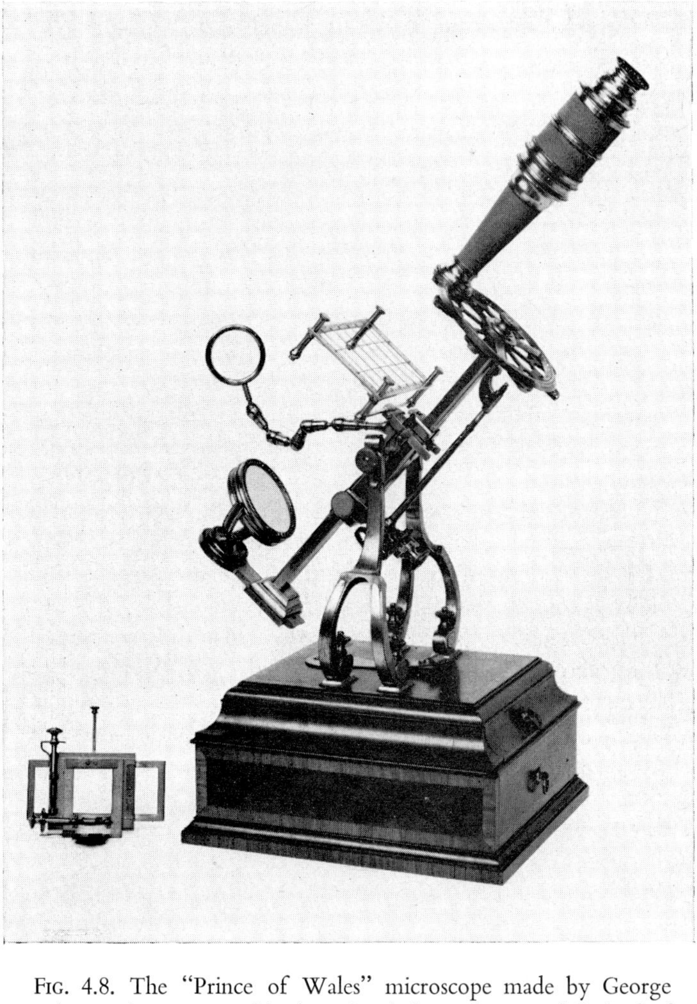

FIG. 4.8. The "Prince of Wales" microscope made by George Adams. The compound body is fitted above the typical "wheel of lenses". The stage fitted to the instrument was intended for holding a frog for the examination of the circulation of the blood. A micrometer stage is illustrated at the left of the microscope.

(*Crown copyright, The Science Museum.*)

Chichester and set up in business as an instrument maker in Fleet Street, London, where he traded from the sign of the "Globe and Visual Glasses". This move to London probably occurred around the year 1750; soon Martin was doing very good business and continuing with his extensive literary activities. Around this period Martin produced many so-called "optical cabinets" which consisted of a drum microscope (developed from his earlier pocket reflecting microscope), a solar microscope and sometimes a telescope, all fitted into a single case. There was obviously a demand for this outfit and Martin, together with other makers, continued to supply ever more and more elaborate versions at least until the 1770's.

In 1759, however, Martin introduced a "Universal" microscope. This particular model may be regarded as one of the milestones in the development of the microscope as it was with this instrument that Martin introduced what was to become known as the "between lens". This represented the first advance in optical design since the field lens had been added to the eyepiece in Hooke's day. The between lens was not Martin's invention, for as he himself pointed out, he found it included in a microscope built at least thirty years before he introduced it into his own instruments. The lens is situated at the lower end of the body, just above the long narrow snout which in those days carried the objective. The extra lens acted to some extent as a compound lens, working with the objective so as to correct some of the spherical aberration of the latter. This allowed the use of larger apertures than had hitherto been possible.

From 1759 onwards, all Martin's instruments were furnished with the "between lens", and were further characterized by the long parallel metal tube fitted to the lower end of the microscope body. This long tube carried the extra lens at its upper end and the objectives were screwed into the lower end. This feature can be seen clearly in Fig. 4.9 (which represents a later development — the "New Universal" model of 1770); it soon became known as "Martin's pipe" and served to carry the Lieberkühn. It is also interesting in that the screw thread which he used to attach it to the body, and to fit the objective to it, was later to become the model from which the Royal Microscopical Society's standard thread was derived.

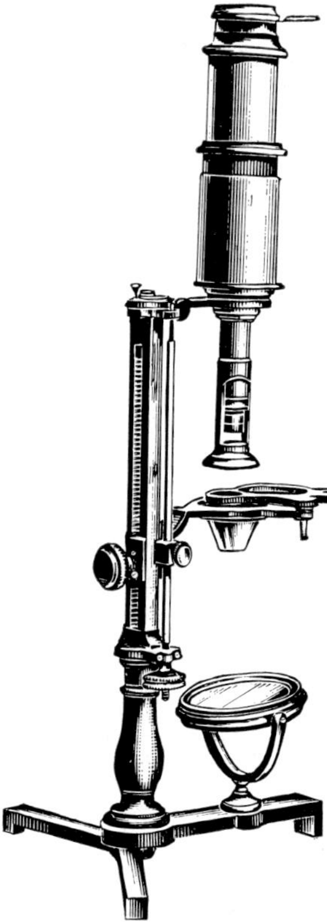

FIG. 4.9. Martin's "New Universal" microscope of 1770. Note the coarse focus rackwork cut into the pillar of the microscope and "Martin's Pipe".

Shortly after Martin's Universal model of 1759 had appeared, George Adams, now instrument maker to King George III, produced his celebrated large silver microscope which he intended for presentation to the King. This instrument (shown in Fig. 4.10) is now preserved in the Museum of the History of Science in Oxford.

This microscope was obviously intended by Adams to represent the height of achievement of the microscope makers of the time; although the resulting instrument may be considered as aesthetically very satisfactory, it does not compare favourably with the earlier microscope which he gave the King when the latter was Prince of Wales. It is quite apparent that in the silver microscope use has been sacrificed to ornament. The stand is constructed of brass and steel cased in beaten silver; there is a large central pillar fluted in the Corinthian style arising from a base which is decorated with cherubs. The microscope actually contains two instruments on the one stand, a simple microscope with its own stage and mirror on one side and on the other a completely separate stage and mirror and the compound body. This latter, as can be seen from Fig. 4.10, is beautifully ornamented and is supported by two female figures. There are eight lenses mounted in a disc, as in the earlier "Universal" microscopes by this maker. The disc can be revolved to change the power in use and so the objectives (which are common to both microscopes) can be brought under either the eye shield of the simple microscope or under the compound body in turn. Provision is made for a substage condenser and also for the mounting of a separate lens to act as a superstage illuminator. All the usual accessories including stage forceps and frog plate are provided.

A careful examination of the actual instrument shows that no really serious observations could be attempted with it on account of the general inconvenience of the construction. If the simple microscope is in use, then the body of the compound microscope is so near that it gets in the way of the observer's head, whilst if the compound body is in use then one has to bend right over the simple microscope. Alternatively, it is possible to move around to the other side of the instrument to avoid this trouble, but then the illumination must be arranged to come from the side. Again there is no provision for the inclination of the instrument so that if it were placed on a table of normal height then the eyepieces are too high for comfortable observation.

At this time, however, microscopes were tending to increase in size and in some of Martin's later models the microscopes were designed to be used on top of their cases which thus served as microscope tables. It is possible that the silver microscope may have been intended for use in this manner, or more likely, it may just have been intended to display objects for the diversion of the King and his guests and used while people were standing around.

FIG. 4.10. The silver microscope made for George III by Adams. The extremely ornate construction is very obvious, especially at the base and in the compound body which is supported by two sculpted figures.

(*Photo: Studio Edmark.*)

Both Martin and Adams continued to produce new models, continuing experimenting with the mechanical parts, for in all essential optical features there was little to choose between any of these instruments of the latter half of the eighteenth century. Martin in particular was a great experimenter and was constantly trying out new ideas on his microscopes, so it is very seldom that any two of his instruments are exactly alike. Numerous variations in points of detail can be seen, many of them probably copied from the productions of his contemporaries. It has already been stressed that all these variations only affected the stand, the lenses (with the exception of Martin's introduction of the "between lens") remaining as imperfect as ever.

His largest and most ambitious instrument appeared around 1780; today it is usually referred to as the "Grand Universal" model and is shown in Fig. 4.11. One of these instruments, formerly in the possession of a famous microscopist of the nineteenth century, Professor Quekett, is now preserved in the Royal Microscopical Society. It must surely be one of the most imposing optical microscopes ever produced, if only from the point of view of sheer size as it stands over two feet high! The general ornate form of the construction can be gathered from the illustration. The limb was of a triangular construction rather reminiscent of that adopted today for optical benches and was inclined from the base by means of a worm and pinion drive. The stage and mirror were both adjustable in position by means of rack and pinions, working on a common rack let into the back of the triangular limb. Full aquatic motion is provided to the body tube (which is now over three inches in diameter) by further rackwork and by a worm wheel. The former controls the distance of the body from the axis of the limb, whilst the latter moves the arm carrying the body in a radial arc over the stage. The instrument contains Martin's "between lens" and in the eyepiece there are two plano-convex eye lenses, whose distance apart may be varied by means of another rack and pinion. The draw tube is furnished with rackwork, another feature well in advance of its time, for this did not become a regular addition to the microscope stand until 1887 when E. M. Nelson had it fitted to the draw tube of his large Powell and Lealand No. 1 stand.

The stage of the Grand Universal microscope is a most complex affair, as there are, in fact, two stages mounted upon the single main

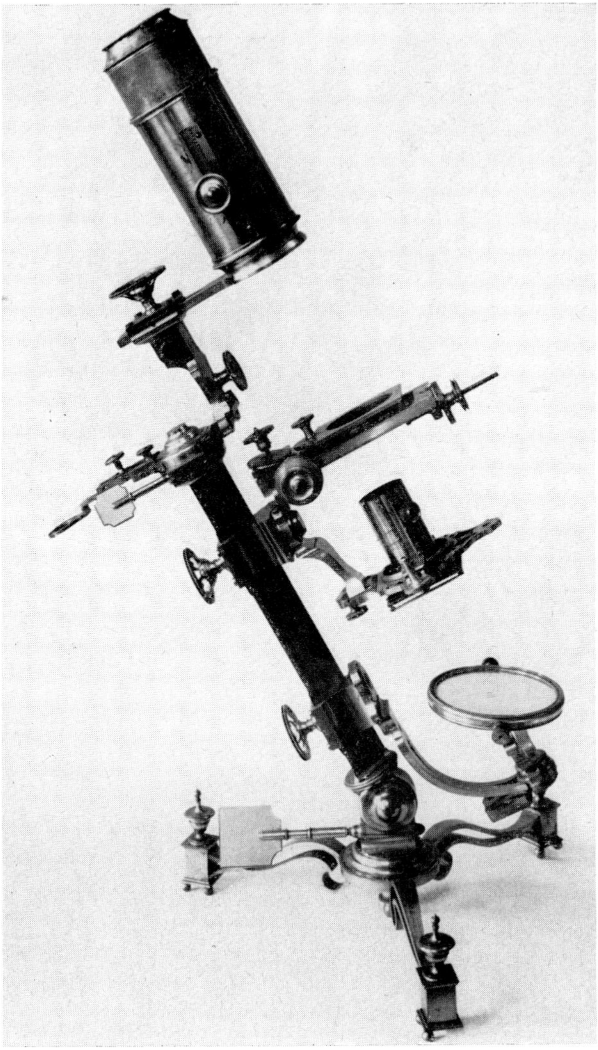

FIG. 4.11. Martin's "Grand Universal Model", now in the collection of the Royal Microscopical Society. Note the triangular pillar, and the various control knobs for the movements. No objective lens is mounted on the body in this illustration.

fitting which has the form of a U-shaped bar pivoting about its centre. By a simple rotation of this fitting either of the two stages can be brought uppermost into the working position. One of these stages carries a tube which serves as a mount for a compound substage condenser, yet another feature in which this instrument is in front of its time. A very large range of objective lenses was provided, one series ranging in focal length from four inches to $\frac{1}{10}$ inch, whilst three lenses of shorter focal length ($\frac{1}{15}$, $\frac{1}{20}$ and $\frac{1}{30}$ inch) appear to be later additions. In addition four lenses intended to serve as magnifiers for the simple microscope are provided. One of the pieces of accessory equipment which is of great interest is a tube which fitted onto the body in place of the snout with the objective. The extra tube contained a concave speculum at its lower end and had an opening in the side through which a pair of stage forceps holding a specimen could be introduced. The object could be brought into focus by varying its distance from the mirror and when illuminated through the hole in the side of the tube a magnified image of the object was reflected to the upper part of the tube where it was viewed by the eyepiece. When this attachment was fitted to the microscope it was in effect converted into a reflecting microscope or "Katadioptric" microscope which is mentioned in Martin's optical writings. The speculum of the instrument in the Royal Microscopical Society's collection has recently been cleaned and now gives remarkably good images, although at a low magnification. Quekett, describing this instrument in his *Treatise on the Use of The Microscope* stated that this microscope

> in point of workmanship, and the extent and variety of its accessory instruments, can probably not be surpassed even in the present day; it is perhaps one of the most complete instruments ever manufactured in this or any other country and serves to show to what perfection the microscope had been brought prior to the year 1780.

This seems a somewhat sweeping statement, for the microscopes which were in production in the middle of the nineteenth century, when Quekett was writing, were superior not only in optical design, but also in the workmanship. The large "Grand Universal" model of Martin, however, must be acknowledged as placing its maker in the first rank of eighteenth-century instrument makers and it provides a fitting tribute to his skill as a craftsman. In addition to its fine construction this microscope is worth noting for the introduction of the triangular limb, a sound engineering feature which was taken up in the next century by such makers as Hugh Powell in his famous instrument made for the Council of the Royal Microscopical Society in 1841 (see Chapter 5). Similarly, the reintroduction of the compound substage condenser, which had not been used since Bonanni showed it in his horizontal microscope of 1691 (Chapter 1, Fig. 1.15), and the rackwork on the draw tube are very significant features which were incorporated generally in later years.

It was largely due to the genius of these two men, George Adams and Benjamin Martin, that the mechanical aspects of the microscope had improved out of all recognition in the latter half of the eighteenth century. There was no systematic development of the microscope, however; Mayall writing in 1886 on the development of the microscope attributed this lack of progress to the fact that the instrument was not being used for systematic scientific studies but was largely in the hands of amateurs and dilettanti who were using it as a means of diversion. Again, the number of microscopes constructed was relatively small and Mayall contended that the serious faults in design would not show up rapidly under these conditions. As communications were slow in the eighteenth century improvements made by one optician would have to be spread largely by verbal or printed descriptions, as probably few makers had the opportunity of examining their rival's instruments in detail. Today with rapid transport and excellent lines of communication new technical developments are very soon exploited throughout the world and progress is correspondingly much more rapid. Again, the design of instruments now proceeds from a sound basis of theory, and the hit-and-miss empiricism of the former days has vanished for ever.

It can be said, however, that by the end of the eighteenth century definite progress had been achieved; the methods of moving the specimen in relation to the optical axis had been perfected and both Martin and Adams were manufacturing mechanical stages with very fine movements. The aquatic motion had been perfected in response to the demands of the observers who were studying the new and fascinating world of the aquatic invertebrates, although the aquatic movement was soon to be abandoned owing to the very rapid wear which soon ensued rendering the stand unfitted for work with high-power lenses. As such lenses were developed, of ever shorter and shorter focal lengths, the need for improved focusing methods appeared. With the development of the rack and pinion the coarse focus was rendered satisfactory and the fine focus system of the Cuff microscope was abandoned until the next development in optics forced the reintroduction of the fine focus, perhaps in a new or much modified form. This process sometimes repeated itself several times. Thus we may regard the second half of this century as a period of intense activity on the part of the instrument makers, although very often the activity and so-called development was haphazard.

Great interest had certainly been aroused among the leisured classes of the eighteenth century by the microscope; this is reflected in most of the contemporary books on microscopy, which devote a large amount of space to descriptions of insects, plants and minerals as they appear under the microscope. With the ordinary compound microscope there is great difficulty in displaying an object to more than one person at a time; even today this problem is relevant to the use of the microscope for teaching, and various accessories have been devised to overcome this drawback. One solution is to have more than one body tube with a prism at the base to direct the light up each tube in turn. This was very popular towards the end of the nineteenth century; one instrument on the market at that time had no fewer than five body tubes each with its ocular, all radiating from a central box containing the prism. The observers sat around a small table and the light from the objective could be directed into each tube in turn so that everyone was able to study the object without too much inconvenience.

The alternative way of arranging for more than one person to view at one time is to use the microscope objective to project a real image upon a screen. This principle is still used in the "conference" type of microscope, although the screen is now usually translucent and the image is projected on it from behind in order to make the accessory as compact as possible. The origins of the projection microscope, however, may be traced back to the sixteenth century "camera obscura", which was used to project an image of the view outside a darkened room upon a paper or screen placed within. This device was much used as an aid to drawing or painting of landscapes. By an extension of this principle the image produced by a simple or compound microscope could be thrown upon a screen, so allowing the object under the microscope to be studied by several people at the same time, and to be drawn accurately, if desired. For all such devices the essential feature is a very strong light source to illuminate the object; in the eighteenth century this effectively meant the use of sunlight and so the use of these projection instruments was restricted to the hours of daylight, and they came to be known as "solar" microscopes.

A typical solar microscope was fixed to a square of wood which was inserted into the shutter of a room. The sunlight was then directed into the microscope by means of a mirror mounted on a circular disc. This could be rotated in a hole cut in the wooden plate by means of a cord which was passed around its rim and around a rod which passed through the plate. This device enabled the whole disc with the mirror to be moved from inside the room. The angle of inclination of the mirror was also adjustable, being controlled by means of a thick brass rod which passed through a cork gland let into the disc and was linked by a jointed arm to the mirror. By a combination of these two movements the sunlight could be directed onto the condensing lens which was fitted in the end of the tube which screws into the centre of the revolving disc. The specimen was mounted in a typical Wilson screw-barrel type of microscope and the image formed by direct projection from the single lens of this instrument.

An example of such an early solar microscope is shown in Fig. 4.12, which is taken from an eighteenth-century woodcut. In this picture the mirror and its hinge are clearly shown as well as the cord drive for the revolving disc. The inclination of the mirror is controlled by the rod H which is linked to the mirror by the joint 7 and the bar numbered 6. The condensing lens is shown at 5 and the tube EDC was for the attachment of the screw-barrel microscope which is not illustrated.

Henry Baker states that such an arrangement would project the image of a louse to a size of 6 feet, "but it is indeed more distinct when not enlarged to above half that Size".

The later solar microscopes of Cuff and of the other later eighteenth-century instrument makers such as Martin, Adams, W. and S. Jones and Dollond, were constructed entirely of brass, and showed a very

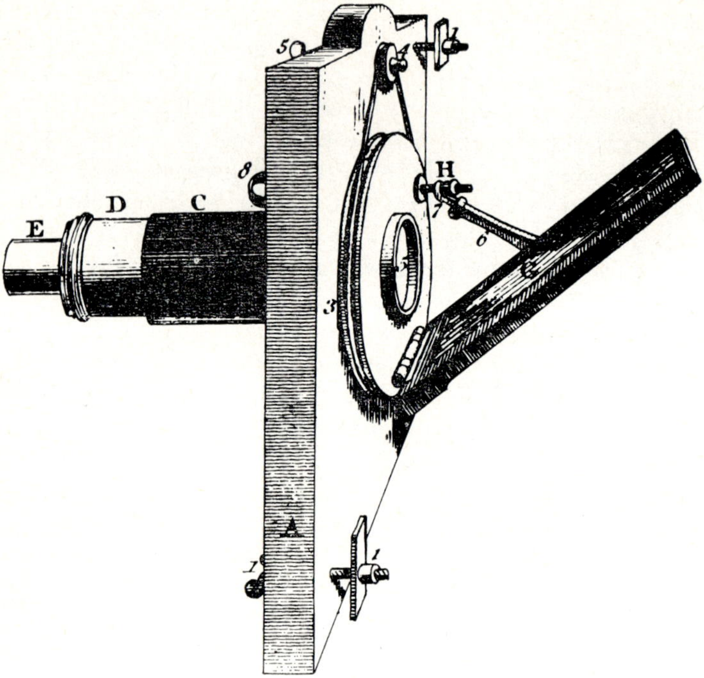

FIG. 4.12. An early wooden solar attachment. The cord and the lever controlling the mirror can be seen to the right of the attachment board (which is labelled A). A simple microscope, probably a screw-barrel instrument, would have been fitted to the tube at E.

high standard of workmanship. Figure 4.13 shows one such instrument by W. and S. Jones, in which the disc carrying the mirror was revolved by means of a rack and pinion, and the angle of the mirror was controlled by the action of an endless screw on a toothed sector. This arrangement (which proved to be much more robust and practical) became universal to all solar microscopes and in this form they continued in production until the middle of the nineteenth century by all the leading British and Continental makers.

This form of solar microscope would, of course, only operate satisfactorily with transparent objects, and as much interest centred

FIG. 4.13. A later brass solar microscope intended for use with transmitted light.

on the study of opaque objects, great efforts were made to adapt the solar microscope for this purpose. This was first accomplished in a practical form by Benjamin Martin, who described his invention in a pamphlet, which he published in 1774, entitled *The Description and Use of an Opake Solar Microscope*. The object was mounted inside a light-tight box attached to the tube of the solar attachment and the sunlight was reflected onto it by means of an inclined mirror. This is seen in the diagram of Martin's instrument which is given as Fig. 4.14. The angle of this mirror could be altered by means of a small screw

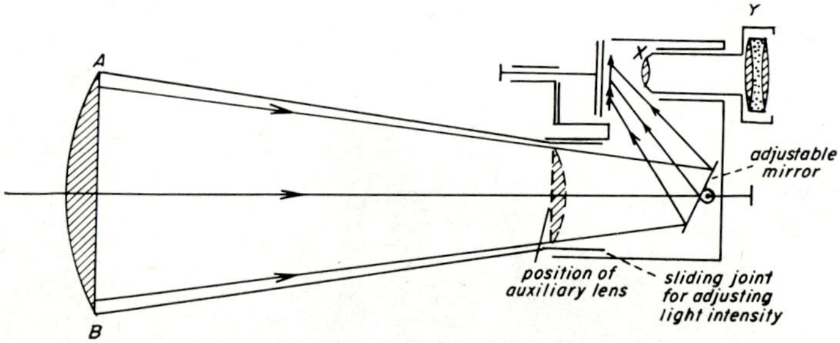

FIG. 4.14. A diagram of Benjamin Martin's "Opake solar" microscope. The solar mirror is not shown. AB is the condensing lens; the object is represented by an arrow and the objective and projector lenses are shown at X and Y respectively. The latter lens is an achromatized triplet.

in order to vary the angle of incidence of the light upon the object. The object-holder was attached to a plate which was maintained in position by means of a strong spring. This slid within a tube and allowed the plate to be withdrawn so that the specimen could be changed. The projection lens system faced the object; in the case of Martin's microscope it consisted of a single bi-convex lens mounted at the end of a tube, together with a large triple achromatic combination at the other end of the tube. Martin said of this instrument:

> With this instrument all opake objects, whether of the animal, vegetable, or mineral kingdom, may be exhibited in great perfection, in all their native beauty; the lights and shades, the prominences and cavities, and all

> the varieties of different hues, teints, and colours, heightened by the reflection of the solar rays condensed upon them.

Under favourable circumstances the solar microscope was capable of providing a most impressive picture upon the screen; according to Adams:

> The effect by this sort of microscope is stupendous, and never fails to excite wonder in an observer at his first view, in seeing a flea, etc., augmented in appearance to SEVEN, EIGHT, or even TEN FEET in length, with all its colours, motions, and animal functions, distinctly and beautifully exhibited.

Later writers, such as Dr. Goring, were sceptical as to the value of the solar microscope. In 1827, Goring wrote:

> We may as well expect to gather figs from brambles, as to get a fine picture from an uncorrected convex lens. The image of a common solar microscope may be considered a mere shadow, fit only to amuse women and children, more especially if we attempt to exhibit brilliant opaque objects, which either become indistinguishable with a limited aperture, or enveloped in a glorious mass of aberration, with an enlarged one. The utmost it can do is to give us the *shadow* of a flea, or a louse as big as a goose or a jackass . . . The swinish vulgar will always be gratified by such spectacles, because they have no idea that a microscope of any kind is to do more than exhibit objects very much dilated in point of bulk; if any optician could contrive an instrument, which would at one swoop take in the whole of a horse, ass, or elephant, and exhibit it (no matter how) with a power of about 240, to this class of observer, I am sure it would delight them infinitely more than the effects of the most beautiful achromatic lenses which Mr. Tulley, or M. Chevalier can make.

From these quotations it seems that in its early days the solar microscope was developed primarily for the display of natural history specimens to several people at one time and that the spectacle was of more importance than the fine detail which was shown. With the development of the achromatic microscope lens the solar microscope soon passed into disuse, helped no doubt by the scathing comments of microscopists such as Goring! One claim to fame does remain to the solar microscope, however, which will ensure that it has a permanent place in the history of the microscope. Throughout the eighteenth century all attempts to recover the microscope image had to be made by some process of drawing from the picture, either freehand or by some form of projection. In the early years of the nineteenth century, however, attempts were being made to utilize the effect of light itself to record the image and these developments led to the photographic process. In this connection the solar microscope was used by J. B. Reade and others around the year 1836 to obtain not only what must be regarded as the first photomicrographs, but pictures which may have a very strong claim to be regarded as the first photographs.

No major discoveries can be claimed for the solar microscope and it remained very much a recreational instrument, depending also for its operation upon the vagaries of the weather. It may well have been for this reason that George Adams the Elder invented an alternative form of instrument which was taken up and improved still further by his son who described it in detail in his *Essays on the Microscope* of 1787. He says:

> This microscope was originally thought of, and in part executed by my father; I have, however, so improved and altered it, both in construction and form, as to render it altogether a different instrument.

The construction of this type of microscope differs radically from the conventional microscope of the time, reverting to the horizontal design first used by Bonanni (see Chapter 1). The lucernal microscope, as it was known, is shown in Fig. 4.15 (redrawn from Adams's book) and in Fig. 4.16 which is of a specimen in the Science Museum in London. This latter instrument has survived with all its extensive accessories intact. From the illustrations it may be seen that the essential construction is that of a large optical bench supported by a pillar and tripod feet; this ensures that all the components are in the same optical axis.

The main body of the microscope is formed by a large pyramidal-shaped box, usually made out of mahogany. This box is about fourteen inches long and six inches square at the larger end. The lenses are screwed onto the smaller end, one of them being shown in position in each of the figures, whilst a pair of large convex lenses is fixed into the larger end of the wooden body. These serve to concentrate the light into the eye of the observer, which has to be placed in a very special position with respect to these two large lenses. In order to achieve this, a "viewing guide" formed from a simple brass ring was placed at the end of the bar which forms the optical bench. When the observer places his eye in this position and looks at the large convex lenses their whole area appears illuminated and contains a very brilliant image of any object placed upon the stage of the microscope.

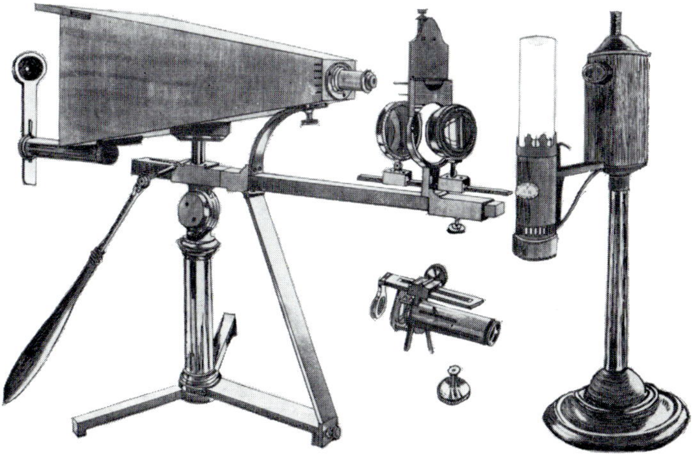

FIG. 4.15. Adams's illustration of his "Lucernal" microscope of 1787. The eye-guide is at the left, and to the right of the mahogany body can be seen mounted the stage for opaque objects. An Argand lamp is shown at the right of the illustration.

It was frequently recommended that transparent specimens should be studied, not by observing the aerial image in the plane of the large convex lenses, but by using the "rough glass" or ground glass to receive a real image. This procedure also allowed several people to view the image at the same time, as when the microscope was used in this fashion there was no need to use the "viewing guide".

All these lucernal microscopes were of course still provided with lenses which lacked correction for spherical and chromatic aberrations. Adams himself was well aware of this failing, for he writes concerning the observation of the aerial image of a transparent object:

FIG. 4.16. A lucernal microscope similar to the instrument shown in Fig. 4.15. This specimen is probably of later date and has the lamp mounted on a swinging arm attached to the microscope itself. Seven extra objectives are seen in front of the microscope.

(*Crown copyright, The Science Museum.*)

> You will then see the object in a blaze of light almost too great for the eye, a circumstance that will be found useful in the examination of particular objects; the edges of the object in this will be somewhat coloured, but as it is only used in this full light for occasional purposes, it has been thought better to leave this small imperfection, than by remedying it to sacrifice greater advantages; the more so, as this fault is easily corrected, and a new and interesting view of the object is obtained, by turning the instrument out of the direct rays of light and permitting them to pass through only in an oblique direction . . . .

Although the lucernal microscope was extremely expensive — it cost £21 — large numbers of them were produced by Adams. He later turned to the use of large cylindrical brass bodies, but W. and S. Jones, who carried on the manufacture of this type of microscope, reverted to the use of the mahogany pyramidal box type of body. With the development of the achromatic microscope the lucernals, like the solar microscopes, dropped out of use.

At this time most microscope manufacturers were well aware of the troubles caused by the lack of chromatic correction in their lenses. Adams regrets that:

> the different refrangibility of the rays of light, which frequently causes such deviation from truth in the appearance of things, that many have imagined themselves to have made surprising discoveries, and have communicated them as such to the world; when, in fact, they have been only so many optical deceptions owing to the unequal refraction of the rays.

He mentions the development of achromatic lenses for the telescope but continues:

> how far this invention is applicable to the improvement of microscopes, has not yet been ascertained; and, indeed, from some few trials made, there is reason for supposing they cannot be successfully applied to microscopes with high powers; so that this improvement is yet a desideratum in the construction of microscopes, and they may be considered as being yet far from their ultimate degree of perfection.

After the death of George Adams the Younger in 1795, William and Samuel Jones continued the business and also bought the copyright of Adams' book. They were responsible for the preparation of the second edition which appeared in 1798. In this the two models of microscope manufactured and sold by the Jones brothers are described and illustrated; known as the "Improved" and the "Most Improved" models they became popular microscopes of the early years of the nineteenth century.

The "improved" microscope is shown in Fig. 4.17, from which the superficial resemblance to Cuff's microscope is apparent.

At this time, Cuff's microscope was still being manufactured, but the "Improved" microscope of the Jones brothers was held to possess several important advantages. It had a larger field of view, the stage and the mirror were both movable, and an "aquatic" motion had been added. This instrument, which was claimed by its makers to be the "second best sort of compound microscope", was sold for £6. 6s. 0d. in 1798. At the same time W. and S. Jones were still advertising Cuff's microscope for £5. 15s. 6d., presumably for the conservative buyer who was distrustful of all the new-fangled developments in the microscope.

According to W. and S. Jones, the best microscope was their other model (Fig. 4.18), usually referred to as the "Most Improved" or, to give it its full description, a "most improved compound microscope, being universal in its uses, and forming the single, compound, opake and aquatic microscopes". In the second edition of Adams on the microscope this instrument is figured and introduced in the following terms:

> A person much accustomed to observations by the microscope, will readily discern the several advantages of this instrument over the preceding one. Besides its containing an additional quantity of useful apparatus, it is more commodious and complete for the management while observing, as it may instantly be placed in a vertical, oblique or horizontal situation, turned laterally at the ease of the observer, and the objects viewed by the primary direct light, or reflected as usual, at pleasure.

The microscope differs chiefly from the "Improved" in that the square-section limb is mounted by a compass joint at the top of a sturdy cylindrical pillar which is supported by tripod feet. As in the other model the actual lenses are mounted in a small wheel which may be rotated under the body tube to change the power in use. Although this microscope did not introduce any novel features,

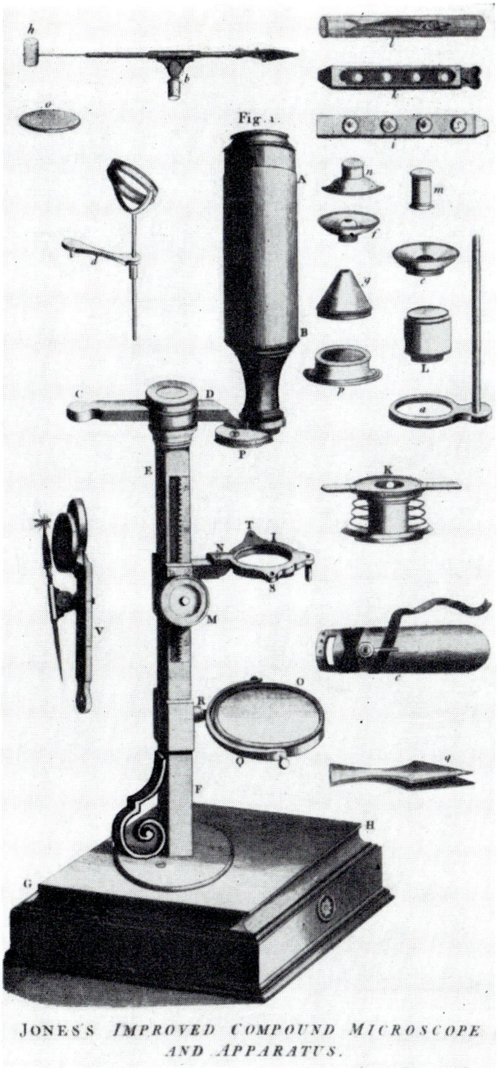

FIG. 4.17. The Jones "Improved" compound microscope. Note the similarity in general design to Cuff's microscope, although later features (particularly from Martin) have been incorporated.

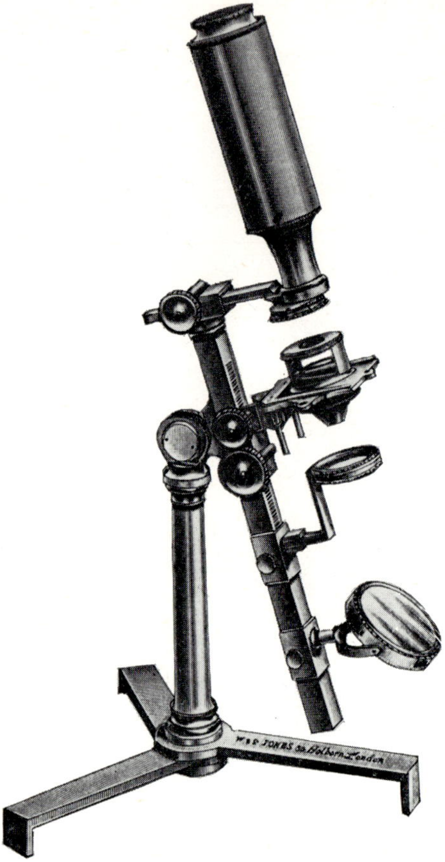

FIG. 4.18. Jones' "Most Improved" compound microscope.

nevertheless it represented an important stage in the development of the instrument as it brought together all the basic ideas which had evolved in the latter part of the eighteenth century, largely as a result of the efforts of Martin and George Adams the Elder. The "Most Improved" microscope was well made and was soon copied by other makers. It performed well, within the limitations of an instrument which did not possess achromatic lenses, and continued in production until somewhere around the middle of the nineteenth century. As made by the Jones brothers it was a fairly expensive microscope, being priced at £10. 10s. 0d. in their 1798 list or £12. 12s. 0d. if purchased with "micrometers and vegetable cuttings", i.e. prepared slides of sections of botanical material.

Throughout the eighteenth century the Continental instrument makers, especially those of France, had not been idle, although few original ideas came from these quarters.

Perhaps the most famous French optician of this period was Dellebarre. He was born in 1726 in Abbeville in France and emigrated to Holland in 1769. He began making microscopes in 1771, whilst he was living in the Hague and soon was persuaded to return to France and set up in business in Paris. Dellebarre proved to be a very successful instrument maker, one of his microscopes being used in the laboratory of Lavoisier, probably the greatest scientist of France at this time. Dellebarre produced many different models of microscopes and when a commission of the French Academy investigated various microscopes in 1777, his instruments received a very good report. Perhaps the most notable feature of his instruments was the large number of separate lenses used in the ocular. It was not uncommon to find four separate bi-convex lenses placed with their surfaces very close together in one of his eyepieces. This may well have been an attempt by Dellebarre to carry out an idea of Euler's that achromatism could be achieved in such a fashion. Mayall, who examined over twenty of Dellebarre's instruments was of the opinion that none of them was achromatic, which is not surprising since the whole concept of Euler's that an achromatic microscope could be developed by using a very complex eyepiece was based on false premises. It was generally held that Dellebarre's instruments were inferior to the best British instruments of this period. Fresnel, a little later when reporting on an early achromatic microscope by Selligue, classified various microscopes by the sharpness or otherwise of their image. He found that Dellebarre's was the worst in this respect, being inferior to the instruments of Adams, Amici and Selligue, whereas in terms of size of field the Adams microscope came out best with Dellebarre's again the worst. It thus seems that at the turn of the eighteenth century the British makers and designers led the world in the production of microscopes, a supremacy which was to be maintained until the later years of the nineteenth century.

Before we pass on to consider the first attempts to develop an achromatic microscope objective, it may be of interest to look at a small microscope which on the grounds of convenience in use, porta-

FIG. 4.19. The Cary type of microscope. This instrument was very small, the total height being about six inches.

bility, and workmanship, may well be regarded as the peak of achievement of the non-achromatic microscope. This instrument was designed in the early years of the nineteenth century by William Cary and it was intended to be simple but at the same time to incorporate all the features which he considered essential for a microscope stand. Portability was high on the list, so the stand in this form of microscope usually consisted of a pillar attached to a round weighted base or to the lid of the instrument case. The mirror was attached to the pillar as was the stage which, although small, often incorporated a simple rackwork movement. This mechanical stage was particularly evident in the later models, dating from around the 1830's. Figure 4.19 represents a Cary microscope which is still in the collection of the Royal Microscopical Society; the small neat construction is especially evident.

There were many variations upon the theme, one of the best known being by C. Gould. Yet another, dating from around 1830, had the

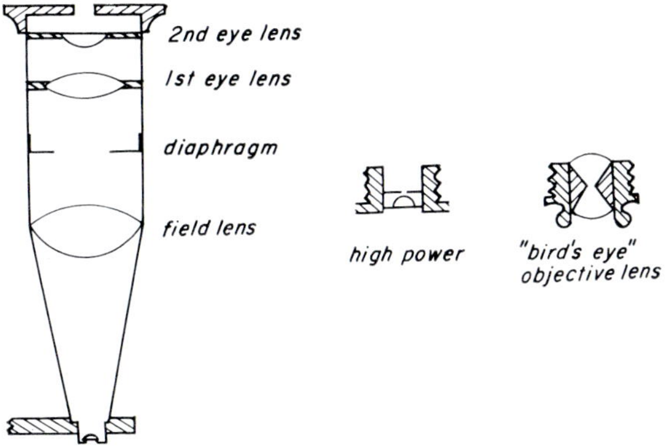

FIG. 4.20. A section of a Cary microscope of about 1830 to show the complex eyepiece and two different types of objective lenses which were used.

optical elements designed by Coddington and they are described and figured in his *Optics* published in that year. From Fig. 4.20 it can be seen that he provided as an objective either a simple plano-convex lens, mounted with the plane side towards the object, and fitted with a diaphragm above the lens to limit the spherical aberration, or for the lower powers a mounted example of one of Coddington's well-known "Bird's-eye" object glasses, which have already been mentioned in Chapter 3. The instrument had a very complex eyepiece, composed of a large field lens mounted at the upper end of the conical portion of the tube and of two eye-lenses, one bi-convex, the other plano-convex. This form of microscope was extremely popular in the early years of the nineteenth century and many examples may still be seen in second-hand shops. The active life of this form of microscope was short, however, as the achromatic microscope became a practical instrument and achromatic lenses required a much larger tube. At the same time the use of ivory sliders to mount the object was declining and glass slips were coming into use; this in turn demanded the development of a larger stage, so that the major design features of the Cary microscope were altered by the radical rethinking of microscope design which occurred around the 1830's.

The great authority which was attached to the writings of Newton had led the majority of the opticians and scientists of the eighteenth century to the mistaken view that chromatic aberration could not be corrected in a series of lenses. Newton's experiments with prisms had proved that refraction was always accompanied by dispersion, i.e. the separation of light into its constituent colours. He thought that the dispersion was always proportional to the deviation of the ray (which is not in fact so) and hence concluded incorrectly that it would not be possible to make a satisfactory telescope by the use of lens systems alone.

Around about the year 1733 a barrister named Chester More Hall, who was an enthusiastic amateur optician and astronomer, began experiments with lenses. He probably worked empirically but succeeded in achieving what the theoreticians believed to be impossible, namely the construction of an achromatic telescope objective. He was aided by the fact that at about this time numerous new types of glass were coming into use, especially the very dense lead-containing glass called "flint" glass. This glass was made in England, but at that time pieces which were large enough to be ground into lenses and were at the same time without flaws were very rare indeed. Chester More Hall evidently stumbled upon the fact that a combination of a convex lens of crown glass allied to a concave one of flint glass would produce the desired result. He appears to have realized the importance of this and the need for secrecy; one component of his lens was sent to Scarlett to be figured whilst the other was placed in the hands of James Mann. It seems, however, that both of these opticians subcontracted the work to the same jobbing optician named George Bass. Much later, when John Dollond became interested in this subject, we are told by Ramsden that Bass passed on to Dollond the details of Chester More Hall's experiments and that Dolland then deduced the purpose of the separate components. There can be no doubt, however, that Chester More Hall succeeded in making an achromatic telescope lens in the 1730's and that he did not take out any patent protection for his invention.

A few years later, in 1755, Klingenstiern the mathematician told Dollond that Newton's hypothesis was not universally applicable; this stimulated Dollond who immediately repeated Newton's experiments and published his findings three years later. He used two prisms, one of glass and the other a hollow shell of glass filled with water. Experiments with this simple apparatus showed that in fact dispersion could occur without refraction and vice versa, so pointing the way to the design of lenses of differing properties which could be combined so that an image could be formed by refraction without the residual dispersion causing colour fringes around the image.

Dollond tried using an idea of Euler's (a Berlin mathematician), that two lenses of glass might have water contained between them to form a third lens. This did not prove very satisfactory in practice because although the chromatic aberration could be controlled, the spherical aberration was very high indeed. It was probably at about this stage that he heard from Bass of the differing properties of the lenses which the latter had worked for Chester More Hall. Dollond tried glass lenses made from these two types of glass which had widely differing dispersive powers. By combining a convex lens of the low dispersion crown glass with a diverging lens of the highly dispersive flint glass, a combination can be achieved which will correct both the chromatic and the spherical aberrations at the same time. In 1759, Dollond used such combinations for the objectives of his telescopes and, moreover, he took out a patent to cover this invention.

Thus by the middle of the eighteenth century the main problem of chromatic aberration had been resolved with reference to the construction of the object glasses of telescopes.

John Dollond and his son Peter pressed on with the manufacture of telescopes based on the principle of combined lenses of crown and flint glass and they became extremely successful. So much so that a petition was lodged against Peter Dollond in 1766 by the London opticians who were forced to pay a royalty on each lens made according to this principle. This action was eventually decided in favour of the

Dollonds; the summing up of Lord Camden acknowledged Chester More Hall's priority in this matter, but commented that

> It was not the person who locked his invention in his scritoire that ought to profit by a patent for such an invention, but he who brought it forth for the benefit of the public.

It seems that for some time no attempts were made to apply the principles of the achromatic telescope lenses to the microscope although in 1774 Benjamin Martin did use a triple achromatic combination in the projection system of his new "Opake solar" microscope. This lens, however, being for projection could be made of a large diameter; the lack of such lenses in the normal optics of the microscope was obviously not due to lack of interest but to the severe technical problems involved in the grinding and polishing of the very small lenses which constitute microscope objectives.

Technical advances in the optical industry were rapid during the last quarter of the eighteenth century and in addition to increased skills in the actual grinding and polishing of lenses, other important developments include the introduction of the cemented lens. In such a lens the various components are stuck together with a transparent cement, usually canada balsam, so that the air film between the various elements of the lens was abolished. This resulted in a great improvement in the performance of the optical combination. One of the first opticians to adopt this new system of lens construction was Chevalier in France, but it was not until the early years of the nineteenth century that such lenses were in general manufacture.

One of the first true achromatic microscope objectives must be attributed to François Beeldsnyder (1755–1808). Here again, as so often in this subject, one of the exciting new technical developments was due to the work of an amateur. Beeldsnyder was a colonel in the Amsterdam cavalry and a member of the Amsterdam committee of Justice. He acted as a collector of the burial rates for St. Anthony's church in the same city, so that like Leeuwenhoek, he could be classed as a minor civil servant. He was an enthusiastic experimental physicist and among his interests was optics, which led him to the problem of achromatism in the microscope objective. The lens which he designed may tentatively be dated about 1791 and it is still preserved in the collections of the

University of Utrecht. The arrangement of the component elements is shown in Fig. 4.21a, from which it is clear that the separate lenses were not cemented together and in fact were widely spaced. They were mounted in a short brass tube. Each of the lenses had a diameter of

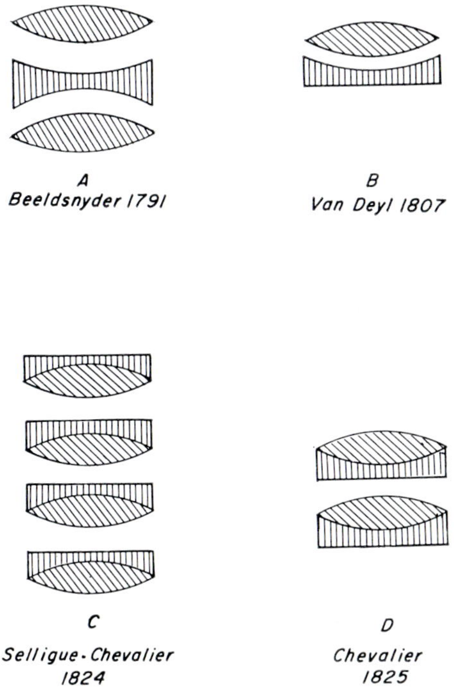

FIG. 4.21. Diagrams of the lens elements in some early achromatic objectives.

about a quarter of an inch and when they were combined they gave an objective with a focal length of just under one inch (21 mm). Van Cittert tested this objective with a Hartnack eyepiece from the same collection and found that the lens had a magnifying power of about 20× and a resolution at this power of 10 microns.

Other pioneers in the construction of achromatic microscope lenses were the van Deyls (sometimes spelt van Deijl). Jan van Deyl

(1715–1801) and his son Harmanus (1738–1809) were reported to have begun the construction of achromatic microscope objectives long before 1791. It appears that they constructed their first achromatic telescope about 1762 and shortly after this attempted the manufacture of microscope lenses on similar principles. Harmanus van Deyl stated that this early lens, which had a focal length of about three quarters of an inch, was entirely satisfactory, but this early lens does not seem to have survived. In the early years of the nineteenth century, Harmanus van Deyl began to sell microscopes with achromatic objectives, but these were probably only produced in relatively small numbers. Several of these instruments have survived and their performance has been studied by van Cittert. These achromatic objectives were composed of a single flint lens together with one bi-convex element of crown glass (Fig. 4.21b). The flint lens was in fact a bi-concave lens, but the curvature of the side which was presented to the object was so shallow that the lens has often been described as plano-concave. This arrangement, apparently hit upon empirically by these workers, would materially assist in the reduction of the spherical aberration in the combination; it anticipates the work of Charles Chevalier (who is usually accredited with the introduction of this disposition of lenses) by about seventeen years. The best resolution which van Cittert could obtain from a lens of van Deyl appears to be 5 microns whilst the focal lengths were either 26 mm or 18 mm. This latter lens, with the highest power of eyepiece which he provided, gave a total magnification of either 92 or 150×, according to whether the draw tube of the instrument was fully closed or fully extended.

In the first decade of the nineteenth century several other workers on the continent of Europe realized that the production of achromatic microscope lenses of low power was now a practical possibility. Bernadino Marzoli, an amateur optician in Italy, was making such lenses between 1808 and 1811. When they were exhibited in Milan in the latter year they were received with acclamation by the Royal Institute of Science. Marzoli's lenses were of advanced construction in that they employed cemented doublets.

Another maker who was producing achromatic lenses at this time was Joseph Frauenhofer. He incorporated achromatic doublets in an instrument which with respect to its mechanical construction was clearly influenced by the "drum" microscopes of Benjamin Martin. Frauenhofer's early achromatic lenses do not seem to have possessed very good definition and it is probable that they were inferior in this respect to the lenses of Harmanus van Deyl.

Other abortive attempts to develop achromatic objectives seem to have been made in France by M. Charles, who does not appear to have been very successful, for Chevalier, writing in 1839, tells us that his lenses were very badly constructed and practically unusable! It is obvious that in the first year of the nineteenth century the imagination of the opticians and instrument makers had been fired by the possibility of constructing workable achromatic lenses of reasonable power. Such a development would have been of the utmost importance for microscopy. It would have helped to raise its standing as a separate discipline and overcome to a certain extent the disrepute into which microscopy had fallen. In the eighteenth century many errors of interpretation of biological structures had arisen, largely because the instruments had a high degree of magnification but a low resolving power and their lenses suffered severely from both spherical and chromatic aberration. There is little wonder that the images which such microscopes produced were so often misinterpreted. Much of this confusion arose as a result of the diffraction haloes which surrounded all the objects due to the very small aperture of the lenses and in consequence it was commonplace for the general appearance of the object to be described as being "globular" or "fibrillar". The "globulist" theories probably reached their zenith in the views and writings of Joseph and Carolus Wenzel and in the works of Milne Edwards who in 1823 reported globules in connective tissue, muscle, tendon, skin and many other tissues. It is interesting that most of the globules which he reported were about three microns in size. As we are accustomed to our vastly improved microscopes of today, and what is equally important, to our carefully controlled techniques of preparing the specimens for examination with the microscope, it is difficult to assess exactly what Milne Edwards was seeing. His results were not confirmed, however, by Hodgkin and Lister who in 1827 were unable to find globules but only fibres in preparations of striated muscle. They were using Lister's new microscope, which will be described later, and attributed the discrepancies between their results and those of Milne Edwards to the deficiencies of the latter's microscope. The situation is well summed up in the words of Baker who wrote in 1947:

> There can indeed be little doubt that many of the globules reported by the early microscopists were images of minute particles, smaller than any ordinary cells, but surrounded by haloes. The fact that the excesses of the globulists were exposed by Lister's microscope seems significant; for the particular advantage of his instrument was that spherical aberration was corrected and the "ring" appearance round small particles thus reduced.

Bichat, who may be regarded as one of the founders of the science of histology, was especially sceptical of the image produced by the microscope: so much so that he preferred to avoid falling into error and avoided microscopical work entirely. He expressed his reservation in the following phrase written in 1800:

> Quand on regard dans l'obscurité chacun voit à sa manière et suivant qu'il est affecté.

The great problem with the first achromatic lenses was that they could be only made satisfactorily to give low magnifications, and as all the really interesting objects required much higher powers to reveal the details of their structure, prospects seemed bleak. The working of the large curvatures required by high-power achromatic combinations did not yet seem feasible so that about 1823, Selligue, a famous French microscope maker, reverted to the system introduced by Adams in one of his microscopes and increased the power by simply screwing together extra lenses. Whereas Adams had simply added more uncorrected lenses, Selligue, however, intended that each succeeding combination should be itself chromatically corrected. The instrument was in fact built in 1824 by Vincent and Charles Chevalier in Paris. It was provided with four achromatic doublet lenses each of which screwed into the other (Fig. 4.21c). Fresnel the physicist was commissioned to report on this microscope and on comparing it with the best of the non-achromatized microscopes then available, he found that up to a magnification of about 200× the new achromatic instrument was definitely superior. Above this magnification there was no improvement in the definition of the achromatic instrument and Fresnel himself preferred the non-achromatic microscope for long observations because of the greater field of this latter instrument.

At this time most of the optical design and construction was still highly empirical; nowadays it is obvious that the basic premise of making a strong objective by combining more and more weaker lenses, each perfectly corrected in itself, was incorrect. The accumulated spherical aberration proved far more detrimental to the image quality and greatly outweighed the advantages gained by the colour correction. The Chevaliers had made the lenses for this microscope of a diameter of about $\frac{1}{2}$ inch, but they were forced to place a diaphragm with an opening of only $\frac{1}{10}$ inch over the topmost lens in order to keep the spherical aberration within reasonable bounds. This drastic restriction of the aperture of the system inevitably affected the resolution very much indeed. Van Heurck in his well-known book on the microscope comments that although the image was well achromatized, the objective was hardly capable of resolving any fine detail. Scales of *Macroglossa stellatarum* (the Humming Bird Hawk moth) did not show any detail when used as a test object although earlier non-achromatic microscopes were capable of showing the longitudinal lines. J. J. Lister, the father of the celebrated surgeon, in his famous paper in 1830 which provided the solution to this problem of accumulated spherical errors in lens combinations, commented on Selligue's lenses that

> The chromatic aberration was thus in a considerable degree corrected, but the glasses were fixed in their cells with the convex side foremost, which is their worst position, and the spherical error was in consequence enormous, showing itself even through the contracted opening, to which it was necessary on that account to limit them.

Chevalier, who actually made these lenses, seems to have realized that a large amount of the trouble with Selligue's microscope was due to the position of the lenses. He himself in his own microscopes, therefore, retained the same method of construction but mounted the elements with the plane sides towards the object and at the same time slightly shortened the focal length of the combination. This arrangement, which is represented diagrammatically in Fig. 4.21d, would give rise to a considerable improvement in the quality of the image, although Lister commented that his glasses were still

> restricted to apertures too small to show difficult test objects.

Again, E. M. Nelson in his contribution to the seventh edition of Carpenter's book on the microscope, noted with respect to Chevalier's objectives:

> Everything in the history points to happy accident as the principal step in achromatized objectives, and this, with very high probability, applies to the work of Chevalier, for Selligue's attempt was a blunder against the commonplace knowledge of his time.

At about this time (1825) we can discern the beginnings of a more logical approach to the design and construction of microscope objectives. The use of special test objects to enable their performance to be assessed accurately and compared with the productions of other opticians was to some measure responsible for this, as well as the most important work of J. J. Lister which gave a proper understanding of the theoretical principles behind the combination of achromatic doublets without increasing the spherical errors of the whole. The credit for the introduction of test objects into microscopy belongs to Dr. C. R. Goring. Goring, a medical practitioner in Edinburgh, first of all had a microscope constructed by a Mr. Adie of Edinburgh. No expense was spared but the instrument, even after extensive modification, was not completely achromatic. Finally, suspecting the competence of Adie, Goring consulted Tulley of Islington who

> soon convinced me of the impossibility of obtaining achromatism . . . otherwise than by the action of concaves of flint glass.

This was not Tulley's first essay into the construction of achromatic lenses, however, as he had previously tried to make them in 1807. Troughton had then asked him to make some achromatic lenses to act as objectives for the microscopes used to read the graduations on the Greenwich circle in the observatory. Tulley made the lenses, each of a focal length of one inch, which proved to be well corrected for chromatic aberration but retained a great deal of spherical aberration so that they were considered a failure.

Goring now acted as Tulley's sponsor and set him to work with the aim of producing a practical achromatic objective in which the spherical aberration was reduced to acceptable proportions. Goring was a prolific writer and he explains his interest in the development of the compound microscope in the following passage:

> I was always addicted to the use of single magnifiers, in my microscopical studies, which I felt to be far superior to all other microscopes in use; yet I always had a lurking fondness for compound instruments, with whose large field of view, and facilities for illumination of opaque objects, as well as the application of micrometers and graphic eyepieces etc., I could not help being much smitten; though I was fully sensible of their great imperfections in point of perfect vision; I thought it would be a glorious improvement if they could be rendered equal in point of optical performance to single microscopes, because they would then have a most decided superiority over the latter, from the accommodations they afford for the observation of every species of object. The eyes of very few individuals are, by nature, gifted with such energy as to endure the use of deep single lenses with impunity for any length of time; and even the most favoured observers must allow that these show so small a portion of an object, and are so disagreeable to use from their very minute aperture, that it is a downright labour to employ them, even when we have been habituated to them for years.

In another passage Goring explains his dissatisfaction with the usual type of achromatic lens current at this time:

> It is but a small point gained to render these lenses free from colour, for they may, notwithstanding, be no better, *or even a great deal worse*, with regard to distinctness, than common ones, as is the case also with the chromatic object glasses of telescopes.

Tulley accepted Goring's commission and set to work. He encountered numerous difficulties and, according to Goring, the

> repeated failures and disappointments which occurred were so very disheartening, that the project was nearly abandoned more than once, not as a physical impossibility in itself, but merely as impracticable in point of execution.

Tulley developed his lenses entirely by trial and error methods using Goring's test object — the scales of the butterfly *Morpho menelaus* — as the standard by which the performance of the various experimental lenses could be judged. This use of a standard object was of vital importance in stimulating the subsequent improvement in lens performance and the development of better methods of lens manufacture. The use of natural test objects was much extended in subsequent years by the use of insect scales which possessed even finer markings, and by the introduction of diatom frustules with their periodic patterning. Such natural objects, however, suffer from the drawback that they are slightly variable so that results obtained with their aid are often not strictly comparable. In the later years of the nineteenth century the test objects which were most used were in fact artificial; Friedrich Adolf Nobert of Barth in Pomerania discovered the secret of producing very regular rulings on glass slides and these rulings, of known periodicity, served as reproducible tests of resolution.

Goring was very much to the point when he wrote in 1829:

> The discovery of a set of objects for ascertaining the defining and penetrating powers of microscopes, has founded a new era in the history of those instruments.

Tulley and Goring proceeded empirically with the development of the lens with the result that Goring was not able to develop the theory, a fact which he readily admits.

> As these small aplanatics have been worked entirely by trial upon new microscopic objects, exactly as a telescopic object glass is worked upon distant ones, very little can be said about their theory; experience has shown that the thinner the component lenses can be made and the more closely curves compacted together, the better.

Tulley produced two partially successful combinations for Goring, which had focal lengths of a third and a fifth of an inch respectively. Later he was able to produce a triplet lens of just under an inch focus with an acceptance angle of 18°.

Goring gave full credit for these achromatic objectives to Tulley, but in his characteristic fashion he contrived to promote his own part in the enterprise.

> The honour of the discovery and execution of its curves belongs entirely to Mr. W. Tulley: yet the public will, I hope, pardon my egotism when I assert that, without my agency, it would never have been made; for at my instigation, and at my expense, was this valuable present to the observer of nature produced.

About this time, Goring began to transfer his patronage to Dollond. Apparently, Dollond was a better workman and produced lenses of a much higher standard of finish than did Tulley. Dollond's lenses were held in their mounts by being burnished into their cells, whereas Tulley's were only screwed in and so on.

Dollond was now producing achromatic lenses, probably of triplet construction, of less than one inch focal length. Most of these lenses were probably not of cemented construction, but retained a layer of air between their components, whereas the Continental practice had been for some time to cement the elements together with canada balsam.

In general, it may be stated that the decade of the 1820's resulted in the production of practical objectives in which the chromatic aberration had been corrected by the addition of concave components of flint glass. Such doublets could then be combined to increase the power of the lens, but the resultant spherical aberration could only be minimized by careful selection and arrangement of the positions of the lens elements by trial and error. No theoretical treatment of this subject had yet been evolved.

The great problem of the residual spherical aberration in compound lenses formed from achromatic doublets and triplets was finally solved by the important work of J. J. Lister which was based on sound theoretical grounds. Instead of empirically correcting the errors of one lens by opposing those of another, Lister showed how to combine fully corrected achromatic doublets so that no further errors were introduced by subsequent elements into the system as a whole. His paper, published in 1830 in the *Philosophical Transactions* proved to be the turning point in the design of microscope objectives; indeed, it may be said that as a result of this work the *design* of microscope objectives became possible, whereas before they could only be constructed on an empirical basis. The starting point in Lister's investigations was the assumption that the microscope objective should consist of a plano-concave element of flint glass cemented to a convex lens of crown glass. For such a lens, shown in Fig. 4.22, Lister deduced that it would possess two focal points (f and f_1) on its axis for which points both the chromatic and spherical aberration would be corrected. For rays emanating anywhere in the space between or beyond these two points these corrections would no longer hold. For rays originating in the space between these foci the combination would be chromatically corrected but *over* corrected for spherical aberration, whereas for the rays which arose beyond these foci there would be a spherical *under* correction. It is not possible here to go into the optical theory behind this, which may be found by those interested in any standard text-book; it is sufficient to note in Lister's own words:

> These facts have been established by careful experiment.

He goes into considerable detail in his paper, which must rank as one of the most important ever published on microscopy and concludes:

> Of the several purposes to which the particulars just given seem applicable, I must at present confine myself to the most obvious one. They

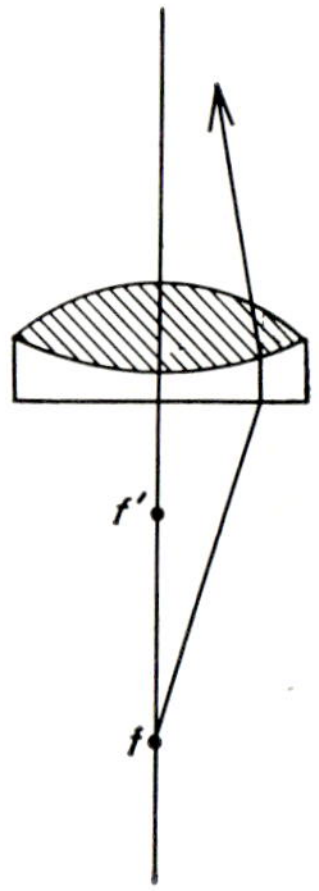

FIG. 4.22. The two aplanatic foci, *f* and *f*′, of a doublet lens. Spherical and chromatic aberrations would only be properly corrected for rays originating from these two points.

> furnish the means of destroying with the utmost ease both aberrations in a large focal pencil, and of thus surmounting what has been hitherto the chief obstacle to the perfection of the microscope. And when it is considered that the curves of its diminutive object-glasses have required to be at least as exactly proportioned as those of a large telescope, to give the image of a bright point equally sharp and colourless, and that any change made to correct one aberration was liable to disturb the other, some idea may be formed of what the amount of that obstacle must have been. It will, however, be evident, that if any object-glass is but made achromatic,

> with its lenses truly worked and cemented so that their axes coincide, it may with certainty be connected with another possessing the same requisites and of suitable focus, so that the combination shall be free from spherical error also in the centre of its field. For this the rays have only to be received by the front glass B from its shorter aplanatic focus f'', and transmitted in the direction of the longer correct pencil fA of the other glass A.

It thus follows that the two completely corrected combinations could be placed at such a distance apart that the resultant rays are entirely free from both chromatic and spherical aberration. The great significance of this discovery, perhaps the most important single feature of Lister's work, was that this distance could be calculated and lens designers were no longer dependent upon a lucky chance arrangement in which the components of their lens happened to be placed at or near this optimal distance.

Lister had Tulley make him some lenses to his own design so that he could experiment further. One of these with a focus of 0·7 of an inch and an angular aperture of 50° gave

> the most distinct microscopic vision that I have yet met with; and I anticipate no serious impediment to the carrying of defining power much further.

Lister found that his ideas did not meet with much support immediately; so little enthusiasm was shown that he decided to learn how to grind and polish lenses himself in order that he could proceed with his experimental work. He achieved a great measure of success in this field, so much so that one of his lenses, which he made to his own design, had the very short working distance of a tenth of an inch and was rated by him as the best of its time. Several of Lister's trial combinations and early efforts at lens grinding are still preserved in the collection of the Royal Microscopical Society. Lister was somewhat surprised that his work led to no rapid development of the microscope objective, and for some years the lens makers carried on in the old way, by means of trial and error. Lister wrote (in a paper which was not, however, published during his lifetime)

> After succeeding fairly well in a trial combination, with this view I left the subject for a while, hoping it would be pursued by opticians but the

> glasses produced by the makers continued to be on the first simple construction of two or three plano-convex compound lenses till the beginning of 1837.

One of the foremost microscope makers at this time was Andrew Ross. In 1831 he began the construction of achromatic lenses and in 1837 Ross produced two lenses to the formulae of J. J. Lister who tells us:

> At that time I called on Andrew Ross . . . and at his request I gave him a projection for an 1/8 inch objective of three compound lenses . . . which he soon worked out successfully, and it became the standard form of high power for many years.

Ross carried on improving the higher powered lens and in 1842 he produced a $\frac{1}{8}$-inch lens with an angular aperture of 74°. These lenses of Ross, produced to the designs of Lister, proved to be of great success. They produced such a good quality image that it was soon discovered that the covering of an object with a very thin plate of glass (or cover-slip as it was known), a technique which was just coming into vogue, was sufficient to reintroduce considerable spherical errors into a well-corrected object lens. The theoretical studies of Lister suggested that this could be eliminated by altering the distance between the lenses which composed the objective, so in 1838 Andrew Ross produced an objective in which the front element was mounted in a tube which was able to slide over a tube of smaller diameter containing the rear element of the lens combination. Screws were provided to clamp the outer tube at any position on the inner one when the desired correction had been obtained. In principle this crude device worked very well but it soon gave way to a screw arrangement in which the outer tube was moved up or down by the rotation of a collar, and so the lens elements were separated or brought closer together.

This device, with very slight modifications, was used until about 1855 when F. H. Wenham produced an improved type of correction collar. In the old version by Ross, the adjustment of the collar caused a movement of the front lens of the objective and unless the greatest of care was taken it was possible to force the front lens of the objective through the coverslip of the actual preparation. Wenham so altered the mechanism that the collar acted upon the rear element of the combination, with the result that the working distance of the whole remained constant. This form of correction collar has survived to this day.

One of the first English achromatic microscopes is shown in Fig. 4.23. It represents the microscope commissioned in 1826 from Tulley by J. J. Lister, and was to be built from the working drawings supplied by Lister. It seems that the actual construction of this instrument was passed on to James Smith by Tulley and it was completed in May 1826.

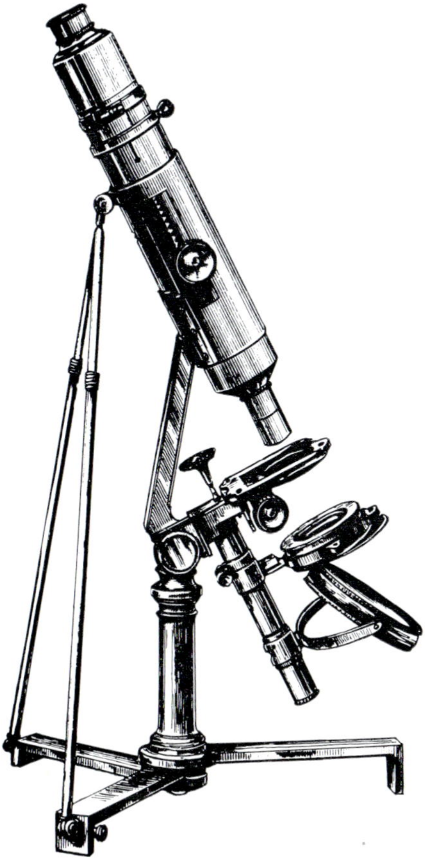

FIG. 4.23. The achromatic microscope constructed by W. Tulley for J. J. Lister.

It shows several features of interest, such as the mounting of the eyepiece in a screwed mount and the provision of full movements to the stage. These movements were controlled from milled knobs, one of which may be seen mounted vertically just in front of the limb; the substage condenser is compound and fits into a sleeve just below the stage.

One of the most important departures from tradition is the fact that this microscope was constructed only as a compound instrument. All the early Marshall and Cuff microscopes were only compound instruments, but the microscopes of Martin, Adams, Jones and Powell up to this time were all intended to serve as both simple or compound instruments. Such was the confidence of Lister in the superiority of his achromatic lenses that he did not include the facility of use as a simple microscope which had hitherto been regarded as essential if fine detail was to be studied. The other feature which is worthy of comment is the device for supporting the limb. Lister realized the importance of stability and steadiness in a microscope stand and he arranged for the body tube to be supported by struts which passed from the tube to the folding tripod base. This feature in a modified form was adopted by other makers in the following years. The workmanship of this microscope had to be of good quality and it is interesting to note in this connection that Lister did not make any provision for the fitting of a fine focus screw, relying entirely upon the smooth running of the main rackwork to focus his relatively high aperture lenses.

A slightly later instrument by Ross which was made by him to the order and to the design of W. Valentine of Nottingham, is shown in Fig. 4.24. This microscope may be dated at about 1831 and to our eye seems rather ungainly; as it could not be inclined, it must have been rather uncomfortable to use when placed on a table of normal height.

The fine focus thread of this instrument is contained within the central stem and acts on the triangular bar which may be seen in the picture. This is an interesting reversion to the practice of Adams in 1746 in which the fine focus milled head is placed below the tripod approximately at table-top level. One of the most interesting features of this microscope is the long tube which served to hold the condensing lens. The mirror was also fitted in this tube at its lower end, opposite

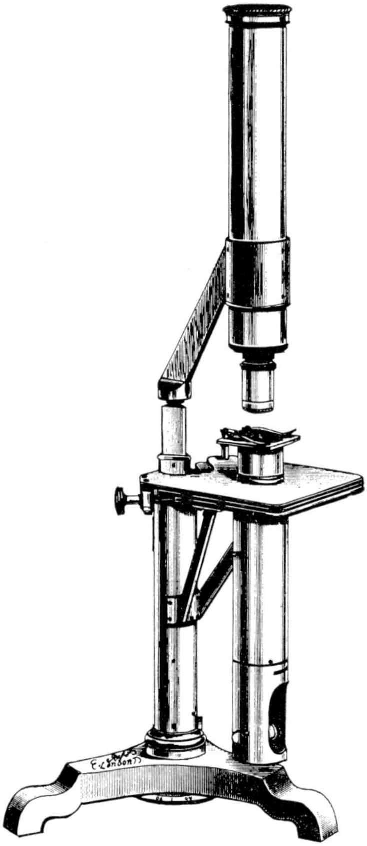

FIG. 4.24. The microscope which Ross constructed to Valentine's design in 1831. The fine-focus control is at table level, below the tripod foot; the screw visible at the level of the stage is one of the controls of the stage movement.

the cut-out portion, and the whole design probably originated with Wollaston who used it in the construction of the stand intended to carry his celebrated doublets. This stand, and the models of a slightly later date to be described in the next chapter, represent the English trends in microscope design. The tripod foot was favoured, together with a long body tube and a mechanical stage in which both the top and bottom plates were movable. On the optical side it was noticeable that nearly all English models were provided with a substage condenser. Allied to these features was superb craftsmanship and a truly fine finish which characterized the best of English instruments throughout the nineteenth century.

CHAPTER 5

The Microscope in Victorian Times

By THE 1830's it was possible to construct a lens in which the chromatic errors had been eliminated and, as a consequence of Lister's work, which was also spherically corrected. This progress meant that objects were now clearly distinguishable under the microscope, whereas before they were blurred and their images surrounded by coloured fringes.

Studies of the microscopic structure of animal and plant tissues were now beginning to flourish, especially on the Continent; as a result of the very clear images furnished by the new achromatic compound microscopes many workers, especially medical men and physiologists, were stimulated to enter the exciting new fields of animal and plant histology. Some microscopists continued to use the simple microscope — one of the most famous was Robert Brown who has already been mentioned in this respect. All his studies on the motion of very small particles in a liquid and on the nuclei in the leaves and staminal hairs of *Tradescantia* were carried out with single lenses of very short focal length.

In general, however, the Continental workers changed to the use of the new achromatic microscope, including Theodor Schwann, who was at this time developing the work of Schleiden. Schwann extended Schleiden's idea of the cell as the unit of plant structure to animal tissues and laid the foundations for Virchow's later studies on cellular pathology which had such important practical applications in medicine.

French instruments were imported into this country and soon accounts appeared in contemporary scientific works of their use in our laboratories and medical schools. This popularity of the Continental instruments among the working scientists was largely due to the fact that they combined good lenses and construction with a very reasonable price. The English microscopes which were being produced at this time were second to none in design and in workmanship but they were large and very elaborate instruments, intended for the

> rich and influential amateurs whose view prevailed in the microscopical societies. These gentlemen, who regarded their costly and monumental instruments as precious toys rather than as the tools for work and study, amused themselves by resolving test objects.

Such microscopes as these were obviously beyond the reach of practising medical men and teachers in the universities, so it is not surprising that they turned to the much cheaper and equally efficient Continental stands by such makers as Oberhauser, Ploessl, and others.

The instruments of the 1830's and 1840's were obviously superior to their predecessors from the pre-achromatic days, but they still had a maximum resolution of only about 1 micron. This figure, which was obtained from measurements made by van Cittert with a Nobert test plate, may be compared with a modern student's microscope of today; on this instrument the $\frac{1}{6}$ inch high-power objective would be easily capable of resolving two points separated by a distance of half a micron.

On September 3rd, 1839, seventeen gentlemen gathered at 50, Wellclose Square, in London to consider

> the propriety of forming a society for the promotion of microscopical investigation, and for the introduction and improvement of the Microscope as a scientific instrument.

Some of the names of those who approved that resolution are now very famous in the history of microscopy in the nineteenth century. Among them were Joseph Lister, George Jackson, Cornelius Varley, Nathaniel Ward, as well as others less well known. As a result of this meeting the "Microscopical Society of London" was formed and a public meeting was arranged. Soon the society could boast 115 members and the famous anatomist Richard Owen was elected to be its first president. At first the society numbered many professional scientists among its members, no fewer than twenty-two being Fellows of the Royal Society, although later the membership was to become largely amateur. This society was later to receive a Royal Charter when it became known as the Royal Microscopical Society, a body which came to rank foremost in microscopical affairs by the turn of the century. Today the Royal Microscopical Society is still active in the field of microscopy, although now catering rather more for the professional worker in this field than for the amateur.

In its early days the amateur members were often "amateurs very near to the border of the professional" as was said of James Scott Bowerbank, one of the founder members and later its president; such men contributed very largely to the scientific knowledge of the day especially in various fields of natural history.

Very often the wealthy amateurs of this country, with their urge to possess the finest equipment possible and to attain the best possible resolution, even if only to score off their rivals, encouraged superb instrument making and optical work by supporting such men as Andrew Ross, James Smith and Hugh Powell. These men were directly responsible for most of the progress in this country around the mid nineteenth century.

Andrew Ross first became noted as a microscope maker when he built an instrument to the design of W. Valentine. This microscope (which was figured in the last chapter, Fig. 4.24) was, like the majority of instruments of its day, capable of being used as both a single or a compound microscope. It resembled Adams's Universal microscope of 1746 in that the pillar was upright and not inclinable; both instruments had flat folding tripod feet and both fitted the screw head for the fine focus mechanism at the base of the pillar. The other interesting feature of this first microscope of Ross was the use of a substage illuminating system which was very similar to that used by Wollaston in his doublet microscope.

The next large stand which Ross produced was illustrated in an article which he wrote for the *Penny Cyclopaedia* of 1839; this microscope is very significant as a landmark in instrument design as the limb had the form of a single casting which supported the body on its upper part and below the stage, on the same casting was mounted the tailstock with the condenser and mirror. This construction is shown diagrammatically in Fig. 5.1, where it is contrasted with the "bar-limb" construction which was the favoured alternative. The former design is generally believed to be the work of Joseph Jackson Lister, in consequence of which it is generally referred to as the "Lister" limb, and it has been extensively used up to the present day.

The Ross microscope of 1839 is shown in Fig. 5.2 from which its general form may be seen. The debt which later microscopes owe to this instrument is clearly obvious. The body tube is supported at two points by a cradle which is driven up and down by the rack and pinion focusing mechanism working on the triangular upper part of the limb casting. The tailstock and mirror mounted on the continuation of this same casting should be noted, as well as the fact that here the cross-section of the limb is now of a circular shape. It is very likely that this microscope was fitted with achromatic lenses, which

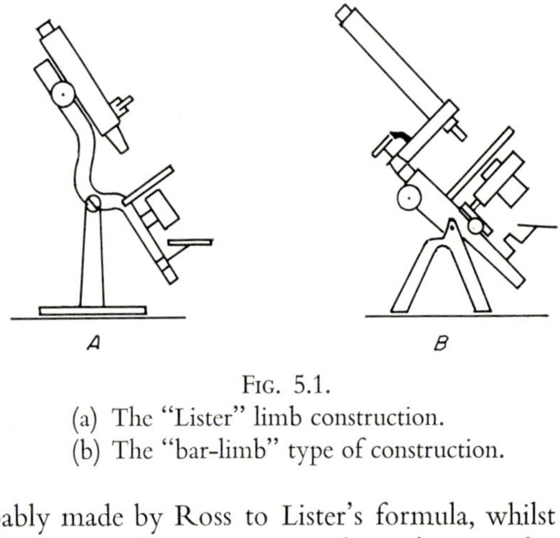

A B

FIG. 5.1.
(a) The "Lister" limb construction.
(b) The "bar-limb" type of construction.

were probably made by Ross to Lister's formula, whilst the higher power lens of $\frac{1}{8}$-inch focus incorporated Ross's coverglass correction collar.

Subsequently Ross adopted another of Jackson's suggestions, which was to mount the limb of the instrument between two trunnions so that the centre of gravity is lowered and a much better balance obtained. An example of such an instrument is shown in Fig. 5.3 dating from 1843 in which the tube is now mounted on the end of a traverse arm which contains the lever of the fine focus mechanism. This form of "bar-limb" model, as it became known, was adopted from Powell and Lealand and was used by most English makers for some considerable time. The microscopes, which Andrew Ross made from this time up to his death in 1859, only differed from this form in minor details, such

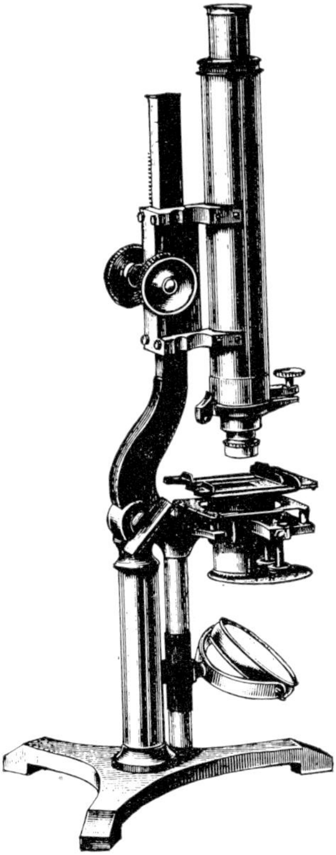

FIG. 5.2. Ross microscope of 1839. The upper part of the limb is of triangular section, while the mirror is mounted on an extension of the same casting, but of circular section.

as the omission of the back support stay for the tube, alteration in the position of the mechanical stage controls, and the provision of a complete substage with rectangular and rotary movements.

In 1841 the Microscopical Society of London ordered sample microscopes from Andrew Ross, Hugh Powell and James Smith. Although the Smith and the Powell instruments are still in the possession of the Society, the Ross microscope was exchanged in 1863 for a binocular model made by Thomas Ross who had continued in his father's business. It is unfortunate that this early microscope is no

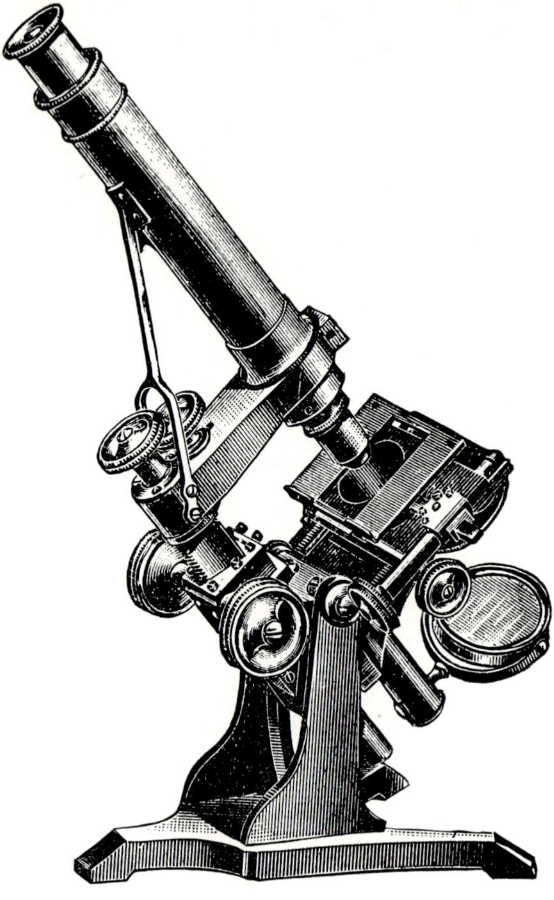

FIG. 5.3. The 1843 model of Ross. Notice the trunnion mounting of the limb, and the adoption of the bar-limb type of construction.

longer extant, but as its delivery was not effected until 1843, it seems very likely that it did not differ in essentials from the instrument shown in Fig. 5.3.

Hugh Powell, like Andrew Ross, was an instrument maker in London. Up to 1840 he had made several microscopes for Cornelius Varley, one of the founder members of the Microscopical Society and a well-known artist. It was not until he was asked by the Society, along with Ross and Smith

> to furnish a standard instrument made according to their own peculiar views,

that he became one of the three leading manufacturers. His microscope was delivered in December 1841 and except for the fitting of a binocular body in 1862, it is still in its original state. This microscope is very solidly built, as can be seen from Fig. 5.4, and it has a system of mounting the body on a triangular bar which is obviously influenced by the Ross microscope of 1839.

Hugh Powell's microscope is fitted with an achromatic condenser, which was almost certainly one of the first ever made in this country, as this was first used in France about three years previously by Chevalier and Felix Dujardin. Powell provided his microscope with a very large solid stage, furnished with a movement (designed by a mechanic named Edmund Turrell) in which the actuating milled heads are concentrically mounted.

At about the time that Hugh Powell constructed his microscope for the Microscopical Society of London, he took into partnership his brother-in-law, P. H. Lealand, and so started one of the most famous associations in the history of the microscope. They continued to manufacture this large model of stand with its characteristic triangular limb; as all of their products are signed and dated it is relatively easy to follow the evolution of their designs. As one of these microscopes is in the Museum of the History of Science at Oxford and bears the date 1849, it seems that this model had a period of production of about ten years.

A few years earlier, however, in 1843, Powell and Lealand had departed completely from this design and produced a stand based on the "bar-limb" model and with a true tripod suspension for the limb. Later changes led to the appearance of the instrument shown in Fig. 5.5, subsequently known as the "P. and L. No. 3" stand. There was the usual Turrell stage, by now standard on all Powell and Lealand microscopes, and a full centring substage fitting.

In 1860, Powell and Lealand brought out what was to become the direct predecessor of their most celebrated microscope, the famous

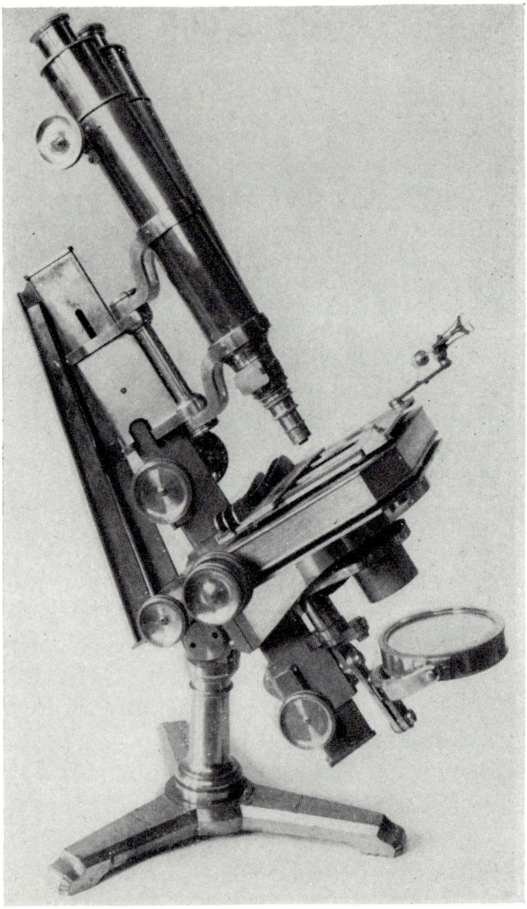

FIG. 5.4. Hugh Powell's microscope, commissioned in 1841 by the Microscopical Society. This particular instrument was fitted with a binocular body in 1862. Notice the triangular limb, obviously influenced by the earlier design of Ross.

"No. 1", which marked the summit of microscope design and manufacture in this country for over fifty years. One of the popular features which was introduced with this model was the provision for unscrewing the monocular body above the bar so that it could be replaced very quickly by a second binocular body. This latter had just come into favour as a result of Wenham's invention of his beam-splitting prism. In 1869 the "P. and L. No. 1" appeared in the form in which it remained for so many years (see frontispiece). The P. and L. instruments were so well constructed that they are still highly prized today not as

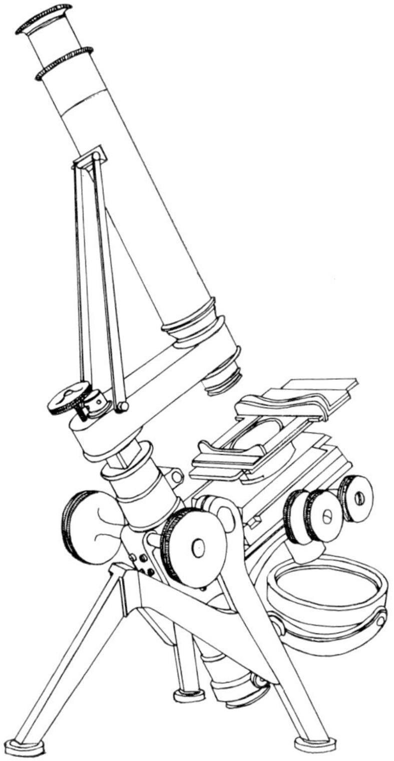

FIG. 5.5. Powell and Lealand's No. 3 microscope. The struts supporting the tube are very characteristic of this model. There is a strong resemblance to the later models of this firm.

collector's items but as actual working microscopes. They can still be used with great success for every type of microscopical research, although naturally they are not so convenient to use as a modern research instrument.

The remaining member of the three great instrument makers of the early Victorian era was James Smith; although he had been making microscopes for the trade in the 1820's (it will be remembered that he actually built Lister's microscope for Tulley in 1826) Smith did not set up in business on his own until around 1840. He too provided one of his microscopes for the Microscopical Society in 1841 and this microscope is illustrated in Fig. 5.6.

In 1846 Smith entered into partnership with Richard Beck and so the famous firm of Smith and Beck came into being; in 1852 Joseph Beck joined the firm, which then became known as Smith, Beck and Beck; large microscope stands were still manufactured on the lines of those already produced, but in order to meet the large and ever-growing demand for a cheap microscope the firm introduced additions to their range. Two of these models are illustrated in Fig. 5.7. At this time great interest was shown in the binocular microscope, as a result of the invention of various types of beam-splitting prisms by Ridell and Wenham. For low powers especially the advantage of using both eyes proved considerable and this type of binocular became very popular. In fact, "Popular" was the name applied by Smith, Beck and Beck to one of their new cheap stands. It was intended to give good service at a reasonable price. This was achieved by simplification of the design and by eliminating many of the over-elaborate refinements which were so common on the large microscope stands of the period. The "Popular" microscope (Fig. 5.7a) used a triangular casting as the base, to which was hinged a broad supporting piece (G) which carried the microscope stage. By placing the tail post (H) in any one of a series of holes in the base, the angle of inclination of the body could be easily changed to any one of several selected positions between vertical and the horizontal. A complete set of accessories, such as dark-field condensers and polarizing apparatus was available for this instrument and could be purchased separately as required.

The other instrument shown in Fig. 5.7b was, as Richard Beck describes it in his book (*The Achromatic Microscope*, 1865)

FIG. 5.6. James Smith's microscope of 1841, commissioned by the Microscopical Society. This instrument has a typical Lister limb and short-lever fine focus acting on the nosepiece.

> the result of an endeavour to make a very low-priced Compound Achromatic microscope by reducing its construction to the simplest possible form, still retaining all that a really useful instrument requires, together with such an arrangement as would admit of considerable additions being made without returning the stand to the maker.

The microscope was mounted on a circular base, the limb itself being carried at the top of a strong pillar, labelled B in the figure. This microscope possessed several interesting mechanical features of design,

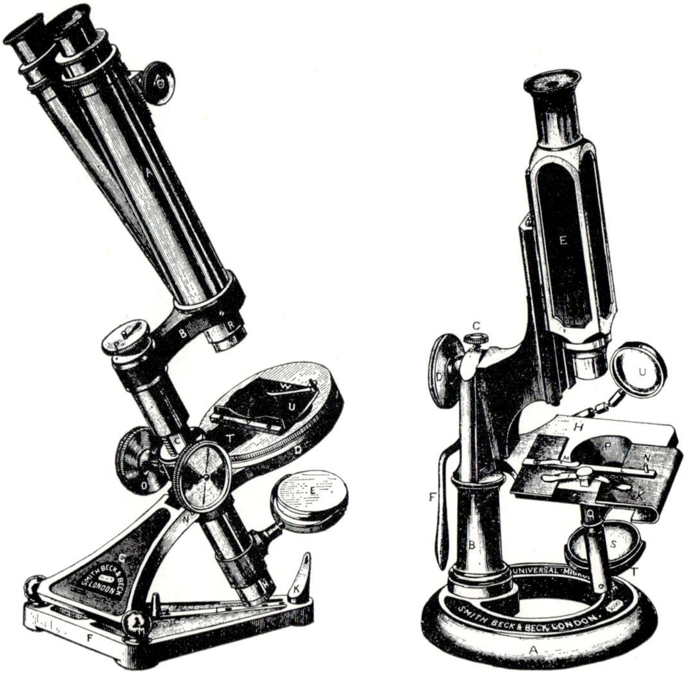

(a) (b)

FIG. 5.7. The Smith and Beck designs intended for wide appeal.
(a) The "Popular" microscope. One of the first cheap binocular microscopes to be placed on the market.
(b) The "Universal" microscope. Notice the square section body tube casting and the fine focus lever, labelled F.

such as the coarse focus control which was mounted on the same axis around which the microscope body pivoted in order to change its inclination. The knob which operated the focus was linked by means of a chain drive to the body E, itself unusual in having a square section, and provided a rapid motion for obtaining the approximate focus of the objective.

Beck's "Universal" microscope also had an unusual fine focus. Hanging free upon the coarse focus knob D was a long lever marked F in the illustration; when this was held at its lower end and pressed sideways, either towards or away from the pillar, it obtained a grip on the axis of the focus knob and allowed it to be turned very slowly indeed. In this way, the need to provide a conventional fine focus (which would of course increase the manufacturing costs very considerably) was avoided. The stage was a simple form of gliding stage which allowed the object to be positioned by sliding the whole with the fingers, after which it was retained in position by means of spring tension.

At the time when this microscope was designed most microscopes were provided with the universal thread for the objective lens screw which had been sponsored by the Microscopical Society in order to promote greater interchangeability between the lenses of different makers. With the "Universal" microscope, however, Beck reverted to a smaller size of thread for the objectives and at the same time set the lenses in as short a mount as possible, so effectively preventing their use on any other instrument. Beck provided five lenses, of focal lengths 2 inches, 1 inch, $\frac{1}{2}$, $\frac{1}{4}$ and $\frac{1}{8}$ inch, together with three eyepieces; these latter were unusual in that they were of the Kellner type which provide a flatter and more extensive field than the normal Huygenian eyepieces.

One novelty which could be added as an accessory to this instrument was the "combined body" (Fig. 5.8). This fitted onto the stand in place of the normal body tube and carried three objectives and three eyepieces permanently mounted on swivelling holders. Only one of each was of course in the optical axis at any one time, but any of the others could be very rapidly brought into use by pressing the head of the pin labelled B in the illustration. The disc C was then free to revolve until the required objective or eyepiece was in position when it was located by the pin springing into the next notch. This accessory was commended to the user in the following words:

> This arrangement may be appreciated by those who are deterred from making a casual use of the microscope, either from the trouble of putting an instrument up, or from the delays which the necessary changes involve, whilst it will considerably assist in the investigation of objects which are undergoing a change either in their position or structure, and when a great range of power is required with the least possible delay.

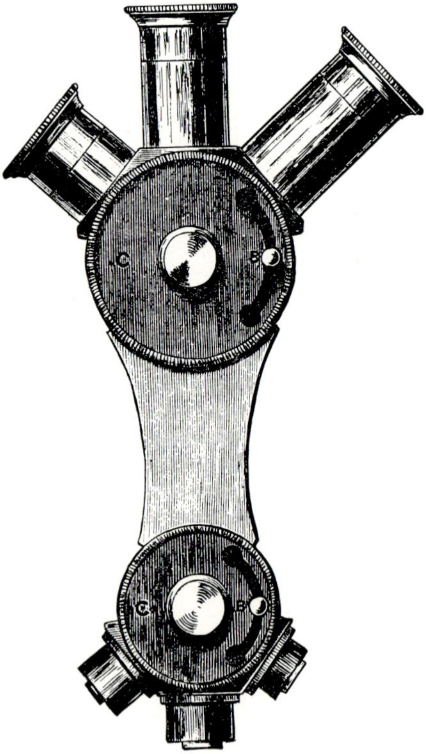

FIG. 5.8. The combined body for the Universal microscope. Any one of the three objectives or eyepieces could be brought into the optical axis by pressing the button on the right hand side of the mounting disc.

In its basic form this microscope retailed for the sum of £9 and considering the quality of the workmanship it must be regarded as a bargain at this price. The provision of such cheap instruments filled a very considerable demand at the time, when there was a growing interest in the microscope which was stimulated by lectures and by the appearance of such periodicals as *Science Gossip* and the *English Mechanic*. The influence which these papers had in the latter half of the nineteenth century on the amateur microscopy of this country has not yet been properly assessed but many of the most eminent microscopists chose to contribute extensively to their pages and many of the controversial issues of the day were hotly argued by correspondence in their columns. Among the microscopists who were active contributors, passing on their considerable technical knowledge were Nelson, Eliot-Merlin, and many others equally eminent who chose to write under such pseudonyms as "F.R.M.S." or "Country Solicitor". It has proved possible in some cases to identify these correspondents and their writings help to build our knowledge of the technical attainments of the period. Frey, writing in 1865 was able to say:

> the use of the microscope has spread widely among the educated section of the public,

and such popularity was on the increase. In an effort to provide suitable microscopes to meet the demand, the Royal Society of Arts organized a competition in 1855 for the design of a microscope to certain specifications. The competition was won by Field and Co. of Birmingham, the model which they produced became known as the "Royal Society of Arts" stand and was sold for three guineas. Most of these stands were unsuitable for really exacting microscopy at high powers, but they were produced and sold in considerable numbers, judging by the number which are still offered for sale by second-hand dealers today! These microscopes must have provided the introduction to microscopy for many people and provided hours of pleasure to their owners.

An example of the microscope stand as it developed on the Continent is shown in Fig. 5.9 which represents the small "No. IV" stand of Reichert. This microscope incorporates the sliding body tube method of coarse focusing and a lever type of fine focus operated by the milled head at the top of the pillar. The horseshoe foot (which

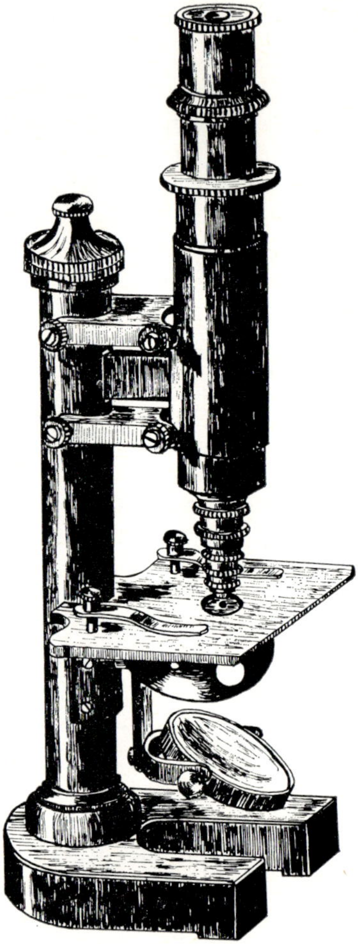

FIG. 5.9. A small Continental stand of the late nineteenth century. There is no substage condenser, some control over the illumination was gained by the use of the wheel of diaphragms to be seen under the stage. Coarse focus is by sliding the whole body tube within a sleeve.

was the popular design on the Continent), the wheel of diaphragms and the concave mirror on its jointed arm can all be seen in the illustration. Towards the end of the nineteenth century, such a microscope, together with three objectives and two oculars would have cost about £6. It was this type of small microscope which was extremely popular with the professional workers. Similar stands were also produced by all the leading makers such as Zeiss, Leitz, Seibert, Nachet and others.

Following upon the researches of Abbe into the theory of the microscope superior lenses became available, which in turn necessitated corresponding improvements in the techniques of illumination and in the mechanics of the stands. We, therefore, find about the year 1880 that the Continental designers moved gradually towards the incorporation of some of the features of English design, although the short tube was still retained. The new lenses of Zeiss, in particular, demanded a very delicate fine focus in order to allow them to operate in a satisfactory manner. In consequence, the new Continental microscopes had to be provided with a much modified fine focus. Again they had to be fitted with a substage to carry Abbe's new illuminator and in order to provide for oblique illumination it then became necessary to add centring screws and a complicated device for moving the iris diaphragm out of the optic axis. Most of these instruments were also fitted with a substantial mechanical stage. Although this late nineteenth century version of the Continental microscope was a very fine microscope (the large stand of Carl Zeiss is shown in Fig. 5.10) many English microscopists refused to accept it; others, however, thought it far surpassed the products of the English makers, and a considerable controversy arose. As usual E. M. Nelson had a very strong opinion which he was not slow to express. Writing in the well-known book *The Microscope and Its Revelations*, part of one of the revisions of which he had helped to prepare, he says:

> The more recent instruments are marvels of ingenuity . . . . There is no fault in the workmanship; it is the best possible. The design only is faulty; there is nothing to command commendation in any part of the model, and seeing that Messrs Zeiss have now progressed so far as to furnish their first class stand with the English mechanical movement, and even stage rotation, and fine adjustment to their newest and best sub-stage condenser, we can

> but believe that the advantages of these improvements will make plain the greater advantage that would accrue from an entirely new model. To all who study carefully the history of the microscope and have used for many years every principal form, it will, we believe, be manifest that the present best stand of the best makers of the Continent is an overburdened instrument.

Despite these criticisms on the part of the majority of the English microscopists the Continental stand, as represented by the Zeiss model Ia (Fig. 5.10) proved to be a tremendous success. Many of the English professional scientists at the universities adopted these Continental models very readily, largely on the grounds that they were used by the best Continental workers who were setting the pace in histology and cytology at this time and so they must be the best instruments for this purpose. At the same time these microscopes were considerably cheaper than the large English stands.

With the development of the Continental model of microscope to its final form, there were thus two completely independent designs current at the end of the nineteenth century, both equally perfect as regards mechanical design and construction.

In the early 1840's in England the three major makers (Powell, Ross and Smith) were producing a wide variety of lenses. In general, it may be said that if we compare English and Continental lenses of approximately the same focal length and of about the same date, then the English opticians were producing lenses of much larger aperture (which, other things being equal, were of correspondingly higher resolving power) than their counterparts on the Continent. The explanation for this may well lie in the class of users for whom the lenses were intended; on the Continent the lenses would have been used by medical research workers who were studying histological preparations.

At this time such preparations would have been largely macerated material, as sectioning techniques which are so widely used today were still in their infancy, and for preparations of whole cells high aperture lenses would have been of no advantage. What these scientists needed was a well-corrected lens which provided good definition at a lower aperture and which provided a large depth of focus or "penetration" as it was then termed. Such lenses could be made in large numbers and at a reasonable price. The wealthy English amateurs on

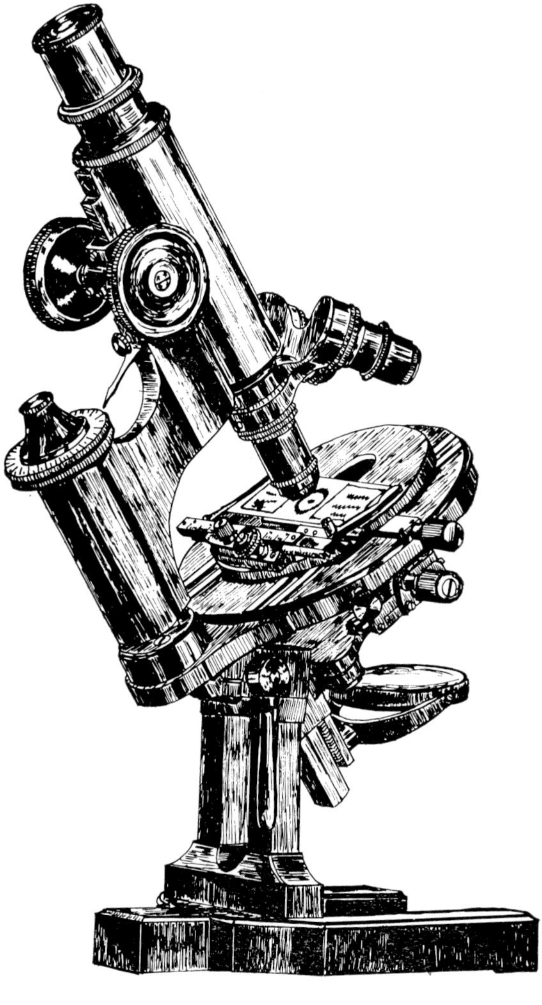

FIG. 5.10. The large Continental research stand in the late nineteenth century. The horseshoe foot is typical of this type of instrument, as also the position of the fine-focus control. This microscope was fitted with Abbe's substage illuminator.

the other hand, who amused themselves by resolving ever finer and finer test objects such as the markings on the frustules of diatoms and the artificial rulings of Nobert, demanded higher and higher apertures. Such lenses which had to be produced by the instrument makers to satisfy their customers, were difficult to make, correspondingly expensive and furthermore they were very tricky to use. In order to exploit their high angular aperture to the full, expensive achromatic condensers were needed and very often as the lenses were of such short focal lengths the objects had to be mounted under very special thin covers.

Despite their interest in attaining high resolving powers, the English microscopists overlooked for many years a very important advance which was widely accepted on the Continent and which was eventually to allow of the attainment of the maximum resolution from the optical microscope. This was the so-called "immersion principle". It was not, of course, new in the nineteenth century; Hooke had used it with simple lenses and he wrote in his Cutlerian lectures of 1679:

> If further, you would have a Microscope with one single refraction, and consequently capable of the greatest clearness and brightness that any one kind of Microscope can possibly be imagined susceptible of, when you have fixt one of these little Globules as I have directed, and spread a little of the liquor upon a piece of Looking-Glass plate, then apply the said plate with the liquor, next to the Globule, and gently move it close to the Globule, till the liquor touch; which done, you will find the liquor presently to adhere to the Globule, and still to adhere to it though you move it back again a little; by which means, this liquor being of a specifique refraction, not much differing from glass, the second refraction is quite taken off, and little or none left but that of the convex side of the Globule next the eye; by which means as much of the inconvenience of refraction as is possible is removed, and that by the easiest and most practicable expedient that can be desired.

It is surprising that this idea was not pursued any further until the nineteenth century.

Immersion lenses were not officially launched until they were exhibited at the Paris Exhibition of 1855 by Amici, who had tried many fluids, such as glycerine, and mixtures of various oils. These latter he did not consider to be practicable at that time as they apparently attacked the cement of the mounts which were in common use. He therefore reverted to water as an immersion medium and his lenses were designed to work with this. They were not provided with a correction collar, however, and so they gave really good spherical correction only when used with a coverglass of a certain definite thickness. This drawback, together with the fact that their resolving power was probably not so high as the best "dry" objectives of the time resulted in the immersion lenses only having a partial success.

It was left to Edmund Hartnack, a very practical optician, to combine the water-immersion principle with the fitting of a correction collar. He began to supply such lenses somewhere around 1859. Hartnack showed that the loss of light increases with the obliquity of the incident rays falling upon the front element of an objective; when objectives of high aperture are used "dry", that is with air intervening between the front lens and the cover slip, the advantages of increased aperture are largely offset by the reflection of a large proportion of the rays falling on the peripheral portion of the lens. If, on the other hand, rays of the same obliquity enter the peripheral portion of the lens from water, then the loss by reflection is greatly reduced and more benefit is gained from the larger aperture. In addition, the use of the immersion principle allows a greater working distance (the distance between the front lens of the objective and the coverglass when the lens is in focus) than would be possible for a dry lens of equal aperture. This is obviously important as it makes the lens much easier to use. Most important of all, however, an immersion lens allows a far greater portion of the light diffracted by the object, which we now know to be important in the formation of the detail in the microscopical image, to enter the lens and so the image quality of an immersion lens is much better than that of a dry lens of equal power.

Hartnack had a great success in Paris with his highly corrected water-immersion lenses. About 1860 his "No. 10" lens of approximately $\frac{1}{16}$-inch focus was produced; this had an angular aperture in air of 172° (corresponding to a numerical aperture of 0·99) whilst three years later his "No. 11" lens had a focal length of $\frac{1}{18}$ inch and an angular aperture of 175°. It was with this lens that Hartnack succeeded in the resolution of the diatom *Surirella gemma* using oblique light but no condenser. He had been experimenting with this lens since 1860 and the early results which had filtered out created such a sensation among the microscopists of the day that they had been putting their names on a waiting list for when the lens should eventually go into commercial production. It is true to say that Hartnack was the first maker to have water-immersion lenses in regular production and from the beginning his lenses achieved a reputation for their very high quality. It was not very long before all the Continental opticians were producing water-immersion lenses; Nobert in Germany was making a $\frac{1}{14}$ and a $\frac{1}{16}$ and later a $\frac{1}{20}$ which could be used either "dry" or immersed by simply adjusting the correction collar. Nachet, Zeiss and Gundlach were also making this type of lens. In the 1860's there was a vogue for lenses with a very high initial magnifying power. This, probably, was due to the special requirements of the English microscopists and it was true both of dry lenses and of immersion systems. Most of the opticians produced the type of lens that their customers demanded and hence lenses of $\frac{1}{25}$, $\frac{1}{30}$ and even $\frac{1}{50}$ inch focus were produced, especially in England. Such lenses with their restricted apertures gave an excessive initial magnification at the expense of image quality and excellence of corrections. Nevertheless, their production was continued for some time until the researches of Abbe showed that their use could not be justified on theoretical grounds.

In England there was great opposition to the use of immersion lenses. Frison, in a recent paper, lays the blame for their rejection squarely on the shoulders of the rich amateurs who by the 1860's were very influential in the microscopical societies and who set the fashion in the microscopical methods. It was probably because the water-immersion lenses could only be used *as such*, i.e. they were not suitable for any other purpose, and because equally good results could be secured by the use of dry lenses employed with better methods of illumination than were in use on the Continent, that the English microscopists did not take up the immersion system. Carpenter, a very eminent microscopist at that time, had himself tried Hartnack's lenses and he agreed that they were capable of showing difficult test objects by means of simple oblique illumination from the mirror alone "with a brightness and definition which are only equalled by the best English objectives when used in combination with condensers and stops". He urged that as the apparatus used in microscopical studies should be as simple and as easy to use as possible, the English workers should adopt this system and demand that the English makers produce immersion lenses at least as good as the Parisian ones.

His pleas fell on deaf ears, however, and even a few years later after Powell and Lealand had begun to make water-immersion lenses and J. J. Woodward in America had used an immersion one-sixteenth of theirs to resolve and photograph the very fine rulings of the nineteenth band of Nobert's test plate, some of the conservative die-hards in the Royal Microscopical Society would not be convinced of the superiority of immersion lenses. The President of the R.M.S. at that time, the Rev. Joseph Reade, who was noted for his pioneering experiments with photography, cast doubts on the value of such lenses by suggesting in his Presidential Address that the

> high praise which is given to the immersion system will perhaps be received with caution until more extended observations have been made upon its powers.

He concludes:

> Still we have yet to learn, notwithstanding these acknowledged advantages whether there are considerable drawbacks looming in the microscopic future which may in some measure counterbalance the employment of water refractions. Of one thing we may be sure, that absolute perfection is unattainable. The ghost of aberration will never be entirely exorcised even by cold water.

Frison attributed the extreme conservatism of the English microscopists largely to the influence of one man — Francis Wenham, who possessed a tremendous practical knowledge of microscopy but only a limited grasp of the theoretical principles involved. He designed many accessories for the microscope, such as a perfected correction collar for objective lenses and perhaps most important of all, his beam-splitting prism for the stereoscopic binocular. Wenham had convinced himself that the immersion principle was a bad one; his views may be summarized by the following quotation from an article which he published in the *Monthly Microscopical Journal* during 1869:

> Recently some excellent glasses have been made, as the so-termed "immersion lenses". These combinations are under-corrected, and not

> suitable for use in any other way. The plan is an old one, and objectives were constructed on this principle by Amici and Ross many years ago. That such lenses give brighter and clearer definition, with the highest powers, from the 1/12th upwards is unquestionable; but the effect of the water and the covering glass is precisely the same in its corrective action *as additional thickness thrown* onto the front lens. The interposition of water doubtless tends to neutralize errors of surface and polish, and more light is transmitted in consequence; but, as preparations are generally not so minute in their structure as to require the habitual employment of such object glasses (which are useless without the water film), it will probably be found, that the inconvenience attached to their use will not compensate for their advantages. Allowing that they perform better on the *Diatoms* and other tests, it would be preferable to mount these objects, in which special discovery is required, without any covering glass, and have an objective of 1/50th constructed with a thick front, specially corrected for uncovered objects only. This having no adjustment, the best definition would always be a matter of certainty, which cannot be the case when the covering-glass, with its varying thickness and errors, forms part of the optical combination.

It seems clear from this extract that Wenham was under a misapprehension as to the reason for the increased light transmission of a water immersion objective, thinking that it was solely due to the overcoming of surface finish irregularities, rather than to the presence of a fluid of higher refractive index than air between the cover and the objective front.

The next logical step in the sequence would be to make the refractive index of the medium equal to that of the glass. At the time of which we are speaking (1860) even the leaders of the optical profession and the most skilful users were ignorant of microscopic theory and so this decisive step was not immediately taken.

In the same year that Wenham was criticizing the use of water-immersion lenses, one of his own countrymen, John Mayall, was making a strong plea for their adoption. It is ironic to note that his article appears in the same volume of the *Monthly Microscopical Journal* as Wenham's criticisms! Mayall was using one of Nobert's new test plates as his object. He writes:

> With the 1/18th and 1/12th by Ross and the 1/20th by Smith, in the possession of this Society, and with a 1/9th, a 1/12th, a 1/16th and a 1/25th

> by Powell and Lealand, all dry objectives (not to mention others which gave similar or inferior results), on a new nineteen band plate, all the bands beyond the 12th seemed imperfect — the lines were not separated.
>
> But with a 1/10th and a 1/18th by Hartnack, of Paris, a 1/16th by Merz, of Munich, and a 1/20th by Nobert, all immersion objectives, straight and well-defined lines were separated as far as the fifteenth band inclusive. In the last four bands true consecutive lines were seen; but they are so extremely slightly ruled, that the eye fails to appreciate their increased fineness.

Mayall concludes that the immersion objectives maintained their superiority by all methods and that

> Continental opticians and men of science have been aware of the merits of the Immersion system during several years past; and to such purpose that knowing how little attention it has received here, they do not scruple to say that the English no longer take the lead either as opticians or microscopists.

Soon after this was written, and perhaps largely as a consequence of it, Powell and Lealand produced their first immersion lens. In an effort to compromise they made it of $\frac{1}{16}$-inch focus and fitted it with a dismountable front lens which enabled it to be used either as a water-immersion or as a dry lens. Soon Ross began to produce immersion objectives, probably at the insistence of Professor Henri van Heurck the famous Belgian diatomist who was a personal friend of Ross. This must have been a bitter blow to Francis Wenham who, according to Frison, was retained as the optical adviser to the firm of Ross and who had so recently been condemning the immersion objective. Although Wenham had great practical skill, he had come to erroneous conclusions in this particular instance and he was later to enter into another even more bitter series of arguments with Robert Tolles over the apertures of the lenses produced by the latter when the new "homogeneous immersion" system was introduced. Again Wenham was proved to be in error, although he never would admit this fact.

Before discussing the introduction of homogeneous or "oil" immersion as it is now called, we must digress for a moment because in 1866 one of the more significant events in the development of the microscope took place. This was the appointment of Ernst Abbe as the technical adviser to the firm of Carl Zeiss. Ernst Abbe, born in

Eisenach, a little town in the Grand Duchy of Saxe-Weimar, was the son of a foreman in a spinning mill. He showed considerable scholastic ability as a boy and eventually attended the University of Jena and later that of Göttingen. It was at this last University that he presented his thesis to obtain his degree. After graduation Abbe taught, first at Frankfurt, and later at Jena University where he was appointed Lecturer in Mathematics, Physics and Astronomy in 1863, when he was only twenty-three years of age. Three years later he was invited to be the optical consultant to Zeiss, who had set up in Jena as an instrument maker in 1846. So began one of the most famous partnerships in the history of the microscope. In 1875, Abbe was made a full partner in the firm and in 1888 after the death of Carl Zeiss he became the sole proprietor.

Abbe abandoned all the preconceived ideas about the design of microscope objectives and set to work to establish the whole process of design and manufacture on a firm basis of theoretical calculations, and to standardize the processes used in the manufacture of microscope lenses. This programme was successful and by 1870 all the lenses which came from Jena were made from standard, pre-calculated designs instead of being produced (as elsewhere) solely by the skill of the individual craftsman. During these early years Abbe carried out numerous experiments and must have acquired a sound grasp of both theoretical and practical optics; it was at this time that the beginnings of the understanding of the influence of the *aperture* of the objective upon resolution (irrespective of its *magnification*) came to him. The results of his studies led him to develop his great theoretical work on the formation of the microscopic image which he published in 1873.

This theory had a great success, despite some opposition, again largely from the English amateurs! It explained how the resolution of fine detail in a microscopic object demanded lenses of large aperture, and why if the structure was periodic (as in a diatom frustule) the use of oblique light with lenses of high aperture appeared to be more effective in obtaining resolution. It is not possible to go into the details of Abbe's theory here, but essentially he showed that a periodic object gives rise to a series of diffracted rays of light on either side of the main illuminating or axial beam. The angular dispersion of these diffracted rays depends on the spacing of the structure of the object; the closer the structure or periodicity, then the wider the separation of the diffracted beams. This is shown in Fig. 5.11. Abbe proved that in order to resolve the structure of the object, the microscope lens must accept not only the axial light rays, but also at least one of these diffracted beams; the more diffracted light which entered the objective, then the more faithful would be the representation of the structure. If the periodicity of an object is so fine that the first of the diffracted beams is so far from the axis that it cannot be admitted to the lens, then there will be

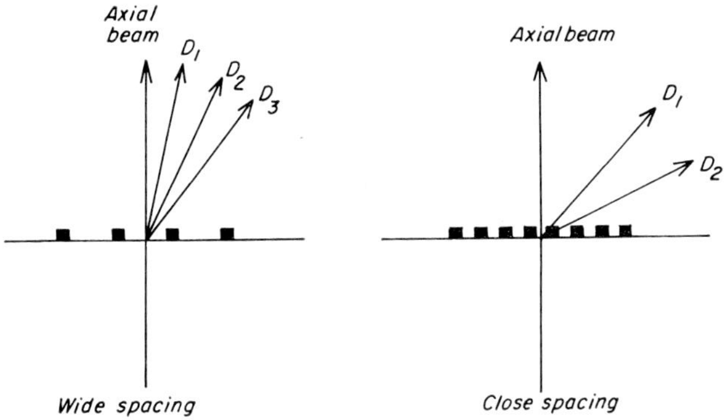

FIG. 5.11. The diffracted beams produced by regularly spaced objects. With a wide spacing the diffracted beams (D_1, D_2, D_3) are close together but with a fine object-spacing the beams are much further apart and separated from the axial beam and each other by a much greater angle.

no resolution of the structure. By the use of oblique light, however, the axial beam is directed into one side of the lens and thus there is then a good chance that at least one of the diffracted beams will enter the aperture of the lens and so allow resolution of the structure to take place. This is shown diagrammatically in Fig. 5.12. This explains the empirical observations that the larger the aperture of a lens, the better was its resolution of fine, regular structure in an object, other things being equal.

It is obvious that by the use of immersion lenses a much larger aperture can be obtained and so a larger amount of diffracted light

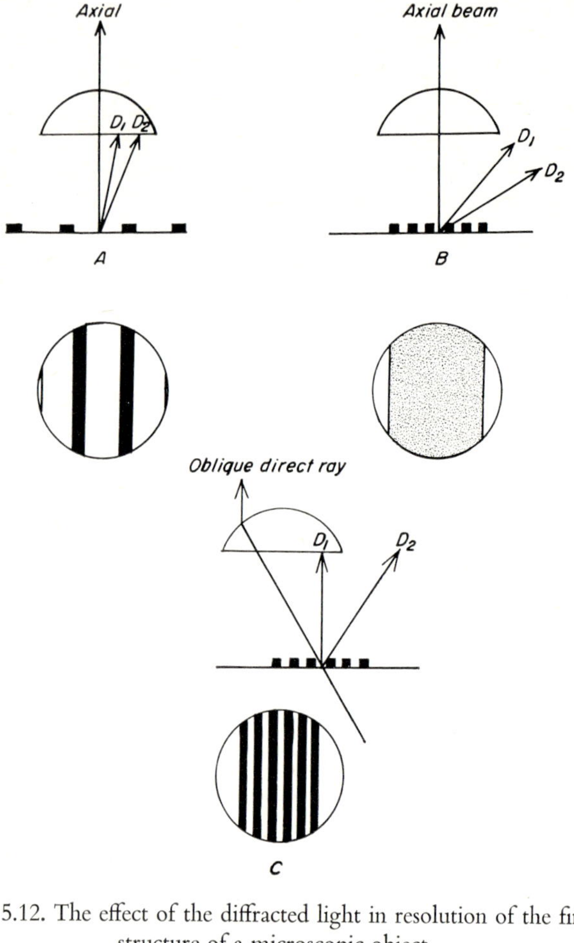

FIG. 5.12. The effect of the diffracted light in resolution of the fine structure of a microscopic object.

(a) Coarse spacing; the diffracted beams D_1 and D_2 are admitted by the objective lens and the grating is resolved.

(b) Fine spacing; no diffracted light enters the lens and no detail is seen in the final image.

(c) By the use of oblique light, the first diffracted beam D_1 is admitted by the lens and the fine grating is now resolved.

can be admitted to the lens, hence the superior resolving power of such lenses. From his theory of image formation, Abbe was led to the concept of expressing the light accepting power of a lens by what he termed the "numerical aperture". Previous to his work there was general agreement that the more oblique the light, the finer the resolvable detail; it was thought, however, that this was directly related to the obliquity of the direct rays of light and that the larger the angular aperture the more suited the lens for accepting such oblique rays. Abbe brought home the point that it was the ability of such high aperture lenses to accept the *diffracted* light, originating at the object, which accounted for their superior resolution. He showed that this relationship could be better expressed not in terms of the angular aperture, measured directly, but to the sine of *half* the angle of acceptance multiplied by the refractive index of the medium between the front lens of the objective and the mount. This quantity Abbe designated the "numerical aperture" or N.A. In the case of a dry lens the expression becomes simply

$$\text{N.A.} = \sin \alpha$$

where the value of α is equal to half the angular aperture, as the refractive index of air is (by definition) equal to 1. This mathematical expression has the great advantage over the older direct form of comparison in terms of degrees of aperture that the resolution of any two lenses may be directly compared simply by comparing their numerical apertures. For example, any lens of an N.A. of 0·8 would be expected to have twice the resolving power of one of N.A. 0·4. Such a direct relationship does not exist if only the angular apertures are compared. One further consequence of Abbe's work (which up to the middle of the 1870's was largely theoretical) was that he was able to develop a formula for calculating the resolving power of the objective so that a theoretical yardstick was then available by which its actual performance could be assessed.

Very soon, however, Abbe was responsible for the practical development of a new advance in microscopy. This was the use of "homogeneous immersion" which was suggested to him by J. W. Stephenson, an eminent English microscopist. Stephenson was interested in obtaining a lens in which the correction collar could be dispensed with; in the *Journal of the Royal Microscopical Society* for 1878 he wrote:

> The origin of the correcting collar as shown by the late Andrew Ross, was the imperative necessity of compensating the error arising from the difference between the refractive index of the covering glass and that of the air between the front of the lens and the thin cover, whenever a high power was used.
>
> Now it is evident that if some fluid, of which the refractive and dispersive powers are the same as those of the covering glass, were substituted for air in the intervening space, the end in view would be attained.

Stephenson lacked the necessary technical knowledge and facilities to carry this any further so he wrote to Abbe suggesting this as a possibility. Abbe followed up the idea and the results were published in his famous paper "On Stephenson's system of homogeneous immersion of microscope objectives", which appeared in the *Journal of the Royal Microscopical Society* in 1879. In this paper Abbe mentions that the same idea had actually occurred to him, but that he was of the opinion that there was no value in it and so he did nothing further with it.

> The idea of realizing the various advantages of such a kind of an immersion lens by constructing objectives in this system had for some time presented itself to my mind, but I thought that there was not much to be expected in regard to the scientific usefulness of such objectives since I believe their use would be limited on account of using oil or some other inconvenient material as the immersion fluid. It appeared to me that, except perhaps for the examination of diatoms, scarcely any other scientific value remained than photographic research which would afford a scope for realizing the optical advantages of such objectives.

Abbe pointed out that not only would the use of homogeneous immersion eliminate all refraction at the front of the lens, but also it would limit the loss of light by reflection. He continued:

> What is still more important, a very considerable amount of spherical aberration is at the same time prevented which otherwise would have to be corrected in the upper portion of the objective.

At the same time the use of homogeneous immersion would have the effect which Stephenson wished to achieve, namely the elimination of the influence of cover glass thickness on the spherical corrections, and hence there would be no need to fit such immersion objectives with a correction collar. The search for the ideal immersion medium gave Abbe a great deal of trouble. He tells us that he personally examined the refractive indices and dispersive powers of over a hundred different liquids, and that Dr. Töpel of Jena studied a further two hundred. Eventually they chose cedar wood oil (to which they sometimes added fennel seed oil to alter the dispersion slightly) and all their objectives were calculated to work with this.

Meanwhile, in America, Robert Tolles had been steadily progressing with the construction of immersion objectives. At first these were water immersion, like all the European lenses, but about 1873 he made a $\frac{1}{10}$-inch lens designed to be used with soft canada balsam as the immersion medium. This was, therefore, a homogeneous-immersion objective as the refractive index of balsam approximates to that of glass.

It thus seems that Tolles deserves the credit for first devising and constructing a workable homogeneous-immersion lens, although his system was rather clumsy and involved a very messy and intractable immersion medium. The achievements of Stephenson and Abbe must be rated equally high, for it is to their work that we owe the practical homogeneous-immersion system, using oil as the liquid medium between the lens and the coverslip. In addition to their practical achievement was added the great theoretical studies of Abbe which defined once and for all the meaning of the aperture of a lens and established its importance in microscopical resolution. Abbe's grasp of the theory prevented any futile controversy from surrounding his discovery and tending towards its disparagement. To a certain extent the achievement of Robert Tolles is clouded by controversy; although he was able and he too understood the theory extremely well, his writings were in such a terse style that his opponents often mistook his meaning and failed to appreciate his point of view.

The new lens of Abbe and Zeiss was keenly studied in England and its performance was described in 1878 by W. H. Dallinger in a letter to the periodical *Nature*. Dallinger, who was very well known for his studies of saprophytic organisms with the microscope, tells us that the lens performed well with the Powell and Lealand condenser or with the

> small plano-convex lens which the maker sends with it to be fastened to the under surface of the slide with the oil of cedarwood.

Dallinger compared it with a new formula $\frac{1}{8}$-inch lens just issued by Powell and Lealand:

> It is but justice to say that all my most crucial tests were equally mastered by the lens of Carl Zeiss. I have not been able to do more with it than the English glass, *but the same results* can be accomplished much more readily. . . . . On the whole, I think it in many senses the finest lens, of its power, that I have ever seen; and in every sense it is an admirable acquisition. . . . . It may perhaps be right to note that this lens, although not provided with the complex arrangement of "screw-collar" adjustment, and although *only* "immersion", is higher in price than the most costly 1/8th by any English maker, although the latter lens may have the screw collar correction, and be both "immersion" and dry.

This new development soon caught the attention of the microscopists and within two years Powell and Lealand themselves were producing homogeneous-immersion lenses; within a very short time an oil-immersion lens became an essential part of the equipment of any microscopist. The professional users of the microscope eagerly sought after these new lenses and it is largely due to the improved images and better resolution of the oil-immersion lens that rapid progress was made in histology in the latter part of the nineteenth century. These lenses were especially valuable in the studies which were carried out in the new science of bacteriology and microbiology; however, we must not forget that the great progress made at this time was not entirely due to these optical advances.

The techniques of preparing the material were also being developed and a technical breakthrough in this field often stimulated a flood of discoveries. One of the main problems with biological material is that tissues and cells are transparent and it is therefore very difficult to discern any structure in the object unless the aperture of the illuminating system is very much reduced so that contrast is provided at the expense of resolution. This problem, of lack of visibility due to lack of contrast, was troublesome until the development of the phase-contrast microscope (see Chapter 6, p. 211) and was first really appreciated when the new high aperture oil-immersion lenses were introduced. The discovery of the aniline dyes was a great step forwards, as it was found that many biological objects could have their visibility in the microscope considerably enhanced by staining with these substances, which rendered them strongly coloured and so made the details of their structure apparent.

The bacteriologists very soon realized the potentialities of the aniline dyes in microtechnique. Robert Koch, who may be considered as one of the founders of modern bacteriology, developed many methods of using these dyes to colour the micro-organisms strongly whilst leaving the background of the tissue unstained. When the high aperture homogeneous-immersion lens was used on such material the occurrence of the bacteria within the tissues was easily seen and their location within the tissues determined. The development of high resolution lenses, together with such methods of differential staining, proved invaluable to bacteriologists and to the cytologists who were using these new methods to elucidate the detailed behaviour of the chromosomes in the cell nucleus at the division of the cell.

At the same time as these existing developments were taking place in the optics of the microscope, equally rapid progress was being made in the techniques of preparing material for microscopical study.

In the eighteenth century it was generally only the surfaces of objects which were examined microscopically, using incident light; sometimes whole objects, if small enough, were dried and mounted dry for examination by transmitted light, but naturally owing to the poor quality of such mounts, very little internal detail could be seen.

One alternative approach is to cut the object into such thin sections that light can easily pass through them. This approach to the study of biological material was not generally possible until some means of supporting the soft tissues whilst they were being cut had been devised. Many workers about 1860 realized the advantages of this method of preparation and a great search started for suitable "embedding" materials which would support the tissue whilst it was cut into thin slices. Frison has recently traced the development of the various agents which have been used for this purpose; the most important was hard paraffin introduced about 1869 by Klebs who melted it around the object to provide the support during cutting. Later several workers adapted the technique to include actually soaking the dried tissue in melted paraffin in order to allow complete penetration of the wax, and so obtaining still better support. This technique revolutionized micro-anatomy as it enables serial sections to be obtained through a complete organism and permits the visualization of the entire structure by a process of reconstruction. Paraffin sectioning is still probably the most used microtechnique today.

The great advances in micro anatomy and cell biology which stem from the period 1870–1900 may thus be attributed on the one hand to the technical advances in the microscope and its lenses, but also on the other hand to the equally important developments in the preparative techniques.

In particular, the bacteriologists made discoveries of great practical significance, as the causative agents of many infectious diseases were soon described. Koch himself discovered the bacillus which causes tuberculosis, which he described in 1882, whilst in the following year he was able to see the agent responsible for cholera. Klebs described the diphtheria bacillus, and Pasteur had characterised the staphylococci and streptococci. In the period immediately after the discovery of the differential staining method by Weigert and a little later after Gram's work on his specific staining technique which enabled the bacteria to be placed into two easily-recognizable categories, more and more organisms were described and the part they played in disease processes was recognized. It may truly be said that the period between 1880 and the turn of the century may be classed as the "golden age of descriptive bacteriology". To this development the new lenses of Abbe were indispensable and together with the development of staining reactions, they may be said to have played a major role in the scientific study of disease.

Most of the medical workers, however, used the oil-immersion lenses without any form of substage condenser thus depriving these lenses of the greater part of their resolution. The view was current at one time that one actually saw *more* if there was no condenser! These faulty conclusions can almost certainly be traced to workers who were studying unstained biological material of low contrast; to see any image at all under these circumstances would have been impossible with a large cone of axial light.

Before the importance of wide aperture lenses was established in the latter part of the century, it had been noticed that the fine periodic structures, such as are found on the frustules of diatoms, were more easily resolved if the light were extremely oblique with respect to the axis of the microscope. This discovery led to the design of some extraordinary microscopes, often of great complexity, in which the substage mechanism was so conceived as to facilitate the production of illuminating rays of great obliquity. All these stands were characterized by a swinging substage, often with the controls elaborately graduated into degrees.

One of the first of these stands was shown by Zentmayer of Philadelphia at the Exposition which was held in that city in 1876. The general pattern of this instrument, including the swinging substage, is shown in Fig. 5.13, which represents an instrument made by Ross to one of Zentmayer's designs. The principle appealed to the Victorian microscopists and later microscopes were made in which not only the substage, but also the object stage was provided with a swinging movement, in order to further the production of oblique illumination. Most of the leading makers produced such instruments and these stands were often of great complexity, every movement being driven by a rack and pinion and provided with a carefully calibrated scale of degrees. Although this complexity does in fact detract from their value as practical microscopes, their construction often showed a very high order of craftsmanship.

Perhaps the extreme form of microscope with such varied movements was reached with the production of the "Ross Radial" microscope to a design of Francis Wenham. The idea behind this instrument (shown in Fig. 5.14 taken from an example in the collection of the Royal Microscopical Society) was to provide complete inclination of the limb and body, together with the substage and mirror system, by sliding it in a sector attached to a rotatable base-plate. This in its turn provided a rotation movement about the optic axis of the whole instrument. There was also a lateral inclination of the limb on either side of the optic axis, so that a wide range of obliquities of the light could be provided in all directions of altitude and azimuth.

With the development and with the spread of an understanding of Abbe's theories, the demand for oblique illumination gradually disappeared; by the end of the century the view was clearly established that for the best resolution a so-called "critical" image was desirable.

This came about largely through the writings and demonstrations of E. M. Nelson, who was an enthusiastic advocate of the use of large aperture lenses used with axial illumination provided by a well-corrected illumination system of an aperture slightly less than that of the objective. Abbe himself provided a simple two lens "illuminator" for use with his lenses; it was first developed in 1873, but was followed

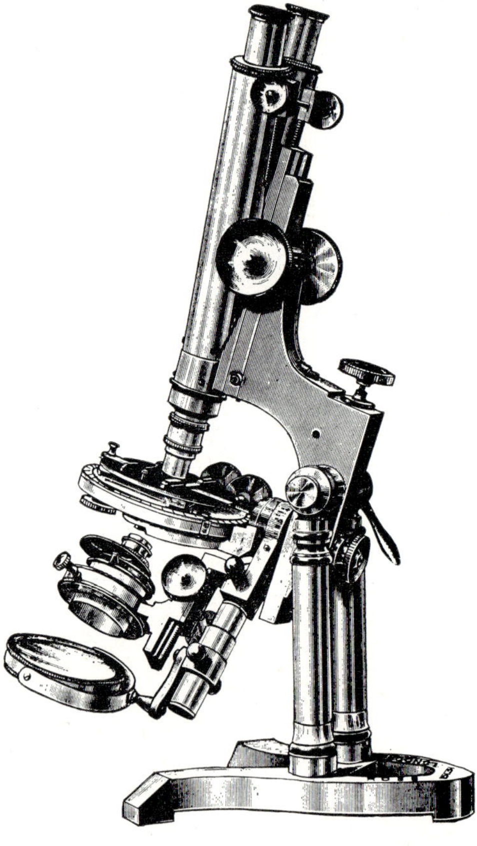

FIG. 5.13. The swinging substage Ross/Zentmayer microscope. The stage could also be inclined in order to increase further the obliquity of the illuminating light.

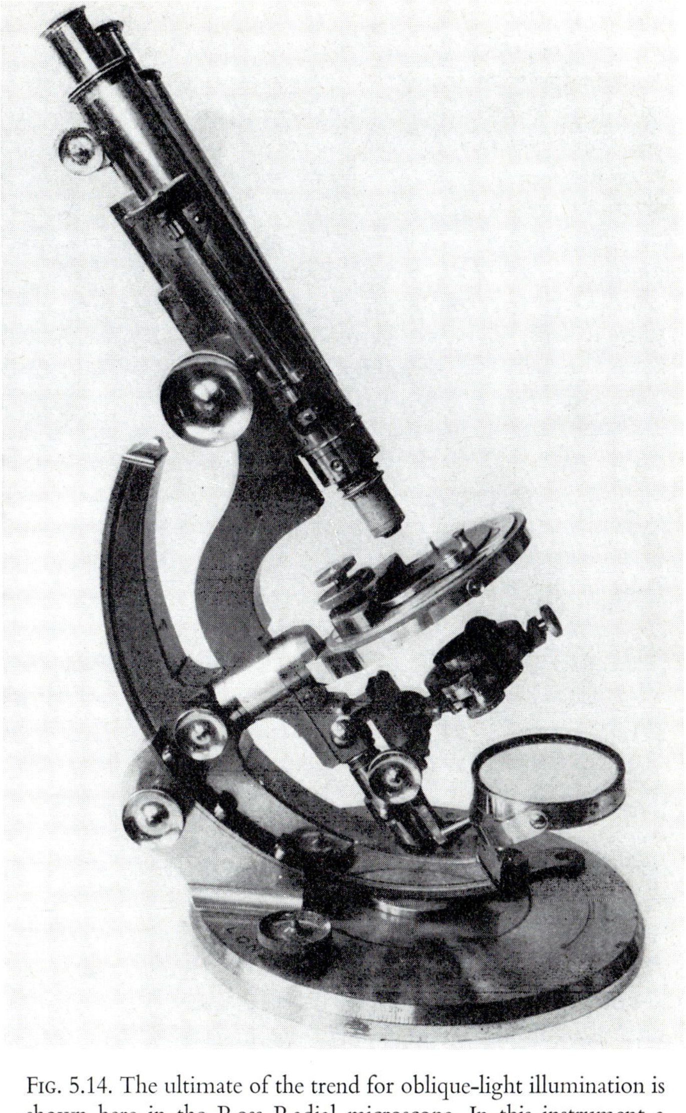

FIG. 5.14. The ultimate of the trend for oblique-light illumination is shown here in the Ross Radial microscope. In this instrument a complete inclination of the limb and body and substage system was possible in all directions.

in 1888 by an achromatic version which was preferable for the more highly corrected lenses. The bacteriologists, especially, soon found that this illuminator gave them a greatly improved image quality and within a few years the Abbe illuminator or condenser as it became known, was the standard fitting for the substage of the Continental microscope. In deference to the still widely-felt need for oblique illumination, Abbe made provision for the iris diaphragm to be moved off the optical axis of the instrument by means of a small rackwork so that the light in its turn became oblique.

It thus can be seen that by the year 1880 the microscope stand had reached the form in which it has essentially remained until very recently, although various bizarre models were still appearing at intervals, and the lenses had been brought to a very high standard by the work of Abbe and the firm of Carl Zeiss. The scientists on the Continent were enthusiastically seizing all the opportunities which their new lenses provided and were rapidly opening up new fields of study for the microscope. In England, once the centre of microscope design and development, conservatism was such that Crisp, writing in 1878 on the "Present condition of microscopy in England", could say:

> In recent years no substantial progress has been made in this country either in the knowledge of the theoretical principles of the microscope itself, or in the systematic investigation of microscopical phenomena.

He concluded:

> It may I think be truly said that out of the entire scientific world there is very probably no body of men who devote so little real attention to the principles that lie at the root of that branch of science of which they are disciples, as do the English microscopists.

These were harsh words, and to a great extent they were well deserved. It may be that the conservatism of the English microscopists (who, banded together in the microscopical societies, exercised such a tight control on the development of the subject) was responsible for this. Certainly the microscopical societies had played a very valuable role in the development of microscopy in England; perhaps at this time they were too conscious of their power and position to view fresh developments objectively. Despite this the British makers were producing instruments of a standard of workmanship unequalled anywhere else in the world. Although the lead in optical design unquestionably lay with Abbe and the Zeiss factory, our own opticians soon were able to produce immersion lenses as good as those of the best Continental makers. Abbe, however, was not content to let his achievements rest with the homogeneous-immersion lens and he sought continually for yet better lenses.

Abbe's formula for the resolution of fine periodic detail in a microscopic object pointed to the importance of aperture. There could be no resolution unless the objective admitted at least some of the diffracted light. The spacing of these diffracted rays depends on the closeness of the object spacing, being much further apart for objects with fine detail; in order therefore to allow diffracted rays from such an object to enter the lens, the aperture must be large. There was thus a very strong tendency to increase the numerical aperture or N.A. of the lenses, as the only other way to increase resolution suggested by Abbe's formula was to use light of shorter wavelength which cannot be carried beyond a certain point.

Powell and Lealand, in particular, soon discovered that it was possible, by making the front element of a microscope objective with a hyper-hemispherical configuration, to increase the N.A. to about 1·43 for an oil-immersion lens of $\frac{1}{12}$-inch focal length. They had some difficulty in mounting such a front element in the metal holder; this they eventually overcame by the simple trick of cementing the actual lens to a plane piece of glass which then served as the actual front of the combination and which could be used to hold the lens firmly in its metal setting. At a later date the same makers had the idea of inverting the whole combination so that instead of making the front element of crown glass and using the flint in the subsequent components, they made the front element of *flint* glass with the crown glass needed to complete the correction of the aberrations in the other combinations. This system allowed the lens to have its aperture increased still further to a value of 1·50, which approaches very closely the theoretical maximum of 1·52 permitted by this method of construction.

It became increasingly apparent that what was needed in a microscope objective was ever-increasing perfection of corrections, allied with increase in the numerical aperture. With the crown and flint glasses which were available to the lens designer of the 1870's and early 1880's it proved impossible to obtain a complete correction of the chromatic aberration. All the coloured rays which formed the image could not be brought to the same focal point because of the fact that the dispersion of the various wavelengths of light by these two different types of glass was not proportional. This meant two colours (usually the extremes, red and blue) were brought together

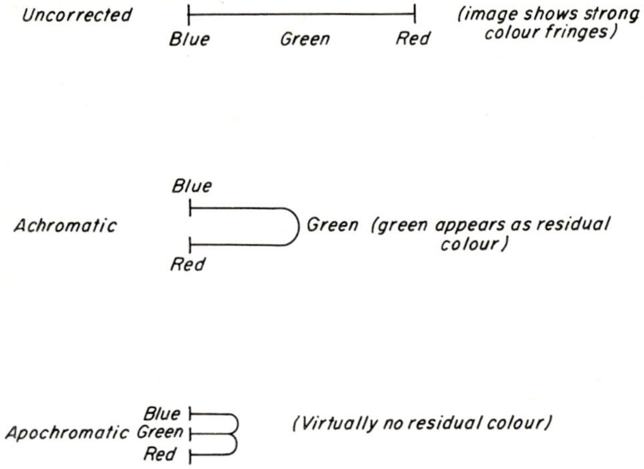

FIG. 5.15. A diagram to illustrate the concept of the "folding over" of the spectrum by an achromatic lens, and the "double folding" by an apochromatic combination.

with the consequence that there was a residual colour in the image. This residual colour was of the wavelengths which form green light so that dark objects still appeared to have a *very slight* border of greenish-yellow light. In an achromatic lens made up of elements of these two types of glass the basic colour correction had, therefore, been effected, but there was still this "secondary spectrum", as it was called, resulting from the folding over of the primary spectrum as shown diagrammatically in Fig. 5.15. A further difficulty was that such correction was only possible for rays passing through one zone of the objective. Again, the spherical aberration in such lenses made only of crown and flint glass could be properly compensated for only one colour. The result was that the achromatic lenses were provided with colour correction for two colours, which were brought together in one zone of the lens, and spherical correction for one colour, usually the residual colour. This gave very satisfactory images, in (say) the central rays but the peripheral rays often produced very pronounced colour fringes at the borders of an object of high contrast, especially if oblique light was used.

Abbe was very sensible of these theoretical restrictions upon the full correction of microscope objectives, and he devoted a great deal of thought to the problem. It seemed impossible to him to correct the errors when the same lens surface would have to be computed to correct both the spherical and the chromatic aberration at the same time. He therefore turned his attention to the possibility of correcting each of these errors independently. This was a completely new approach and required, as he wrote in 1879:

> at least two kinds of glass having optical relations different to those now in use, — either low refractive index combined with high dispersive power or high refraction with low dispersion.

No such glasses were, however, known in the 1870's.

When he visited the Loan Exhibition in England in 1876, Abbe indicated in his report that, from the point of view of the owners of the great glass works, experiments designed towards the creation of new types of glass were not considered practicable; Abbe suggested that the State Academies might be induced to help in the solution of this problem. One immediate result of the publication of this report in Germany was that Otto Schott sought out the acquaintance of Abbe. Schott was the son of a glass maker and fully understood the manufacturing processes involved; at the end of 1879 he wrote to Abbe and told him that he had succeeded in formulating a lithium-based glass which possessed different optical properties from the normal types then in use. Abbe encouraged Schott to persist in his experiments and to develop systematically the manufacture of new varieties of glass; Schott persevered and in 1881 Abbe was able to report to him that two new borate glasses had properties which would make them of use in lens manufacture.

About this time Schott was persuaded to move to Jena and he was established with the aid of financial help from Zeiss in a small research laboratory.

In 1882, Schott had given Abbe small samples of optical glasses with properties which were widely different from those of ordinary crown and flint glass. From these samples Abbe made two lenses, one of which had a focal length of 1 inch and an N.A. of 0·3 whilst the other had a focal length of $\frac{1}{6}$th of an inch and was of 0·86 N.A. The Berlin instrument maker Bamberg, who was closely associated with Carl Zeiss, tells us that the first of these lenses gave a definition

> as has perhaps not been accomplished before combined with great brightness and complete correction of every trace of chromatic aberration.

Abbe himself reported later (1884) with respect to both of these lenses

> with both objectives, that which hitherto could not even be attempted with any similar lens system has now been almost absolutely achieved, namely equality of the spherical aberration for two different colours (i.e. elimination of the chromatic difference of spherical aberration). In the weaker one it has at the same time been possible to completely eliminate the secondary chromatic aberration at the axis, in other words bring the exact union of rays of three colours, and this is probably the first instance where this condition has been satisfied by an optical construction of any kind which has actually been made. In the stronger objective, on the other hand, the uncorrected remnant of the secondary spectrum has been reduced to slightly less than half the amount which must remain at the given angular aperture with other available types of glass.

He tells us that in these lenses a total of not less than seven new types of glass were used.

With the help of a young assistant, Paul Rudolph, Abbe calculated the data for some new lenses and they were announced to the world at a lecture given by Abbe himself to the "Jenaische Gesellschaft für Medizin und Naturwissenschaft". An English translation of his paper soon appeared and was published in the *Journal of the Royal Microscopical Society* for 1886 with the title "On Improvements of the Microscope with the aid of New Kinds of Optical Glass". This paper described in detail the properties of the new lenses, stressing their excellent definition and at the same time describing the special eyepiece which could be used with them to enhance their performance. Abbe called his new lenses "apochromatics" (meaning "away from colour") in order to emphasize the fact that the elimination of the secondary spectrum resulted in

> achromatism of a higher order than has hitherto been attained.

In an apochromatic lens three colours are united for two zones and its spherical aberration is corrected for two colours. As shown diagrammatically in Fig. 5.15, the union of three colours at one focus results in the spectrum being "folded" twice, so that the resultant image is almost entirely free from residual colour.

What practical advantages did the use of these new and necessarily very expensive lenses confer on the user? For the greater part of microscopical studies it must be admitted that the advantage would go unnoticed; for careful researches involving the use of the highest powers of the microscope at the limits of resolution then the image quality would show a definite improvement. With the ordinary achromatic lenses, for example, the working aperture of the lens had always to be reduced, as the light passing through the poorly-corrected outer zones spoilt the quality of the image. Apochromatic lenses, however, allowed the use of their full stated aperture; in addition their corrections were so good that it proved possible to use eyepieces of very high power ("deep" eyepieces as they were called at the time) and so obtain a more easy visualization of all the detail that their resolution afforded. It was also stressed that the almost perfect achromatism of these new lenses would be valuable for photomicrography which was rapidly developing at this time as a means of recording the microscopic image.

Very soon the new lenses were constructed by other makers; Powell and Lealand in England were listing them by the year 1892 and soon Reichert, Leitz and others were making them on the Continent. It is surprising that Abbe did not even trouble to protect his invention by taking out a patent. No sooner had the apochromatic series appeared on the market than Abbe began work on yet another idea, the development of the use of fluorite as a material for the construction of microscope lenses. Fluorite is a natural crystalline material with a much lower refractive index and dispersion than any known optical glass and it had been realized by Abbe about 1881 that this substance possessed optical properties which would make it of use in correcting some of the residual spherical and chromatic aberrations in microscope objectives. In the published descriptions of the new types of apochromats, the stress was laid on the use of Schott's new varieties of optical glass and nothing was said about the use of other compounds such as fluorite, although it is likely that this substance was in fact used in the first apochromatic objectives. Abbe devoted a lot of time and effort to procuring supplies of fluorite and to developing its use in lenses; in 1890 he published detailed information as to its use in combination with the Jena glasses. This proved to be a very important development as nowadays "fluorite" lenses are extensively produced with corrections intermediate between those of an ordinary achromat and the much more expensive apochromats.

The new apochromats were soon in great demand by microscopists, who appreciated the superb quality of image which it was possible to obtain by their use. One difficulty soon became apparent — the new Jena glasses were not very stable, and in hot or humid climates especially they soon began to devitrify and eventually became completely opaque, so making the lens useless. The manufacturers actively pursued research to discover yet more types of glass which would not suffer from this fault. In the meantime, Zeiss issued a new set of lenses (in 1888) which had been recalculated and in which the number of lens elements had in some cases been increased to give superior corrections, especially of the marginal rays. Finally, in 1894, Zeiss began to introduce yet a third system of apochromats, this time manufactured of glasses which were much more durable; by 1897 all the apochromatic lenses were being produced to these improved formulae.

These lenses were much favoured by the bacteriologists, who had been among the first to appreciate the value of the homogeneous-immersion system a few years earlier. Among the many discoveries which were made with the new lenses perhaps the observation of the unstained syphilis organism by Schaudinn may be singled out for mention. By the aid of the very high resolution the *Spirochaeta pallidum* was clearly visualized for the first time. Such was the enthusiasm for these lenses, which of course demanded the most meticulous attention to the arrangement and alignment of the microscope for their successful operation, that some workers wished to use them for any and every purpose, often under the most unsuitable conditions! The microanatomists and the histologists and cytologists also became aware of the enormous potentialities of these new lenses and as an example of the superb results to which they gave rise, one has only to look at the description of the spermatozoa published by the Swedish anatomist Gustav Retzius, which have not been surpassed by any optical microscopist since. With proper use, the apochromatic objective allows the attainment of the ultimate resolution of the optical microscope. By 1890 these lenses were allowing expert microscopists to resolve structures which lay at the theoretical limits of visibility (about $0 \cdot 25\ \mu$). Workers such as E. M. Nelson in this country, van Heurck in Belgium, Retzius in Sweden and others were pushing forwards the limits of the optical microscope. Nelson was very impressed with the new lenses and wrote in 1889:

> Thanks to the Abbe Zeiss apochromatics, the curtain which hides the very small from our gaze has been drawn up a little higher.

One has only to study the catalogues of the leading manufacturers at the end of the nineteenth century in order to see how many superb lenses were produced, often with extremely high apertures and of apochromatic quality. Powell and Lealand, for instance, in 1893 were offering such a lens of $\frac{1}{10}$-inch focus with an N.A. of 1·5, a truly remarkable figure. Such a lens, regarded as extremely fine by van Heurck was, it is true, very expensive, costing £50; at this time such a sum of money would have put this lens beyond the reach of most people. With the aid of such high aperture lenses it was perfectly feasible to push the resolving power to its limit. Booth, in a recent article in the *Journal of the Royal Microscopical Society* comments on this in the following words:

> The amateur microscopists of the nineteenth century were responsible for the great technical virtuosity of the opticians of the time. It is a salutary comment on scientific progress that whereas between 1890 and 1914 one could buy either from Charles Spencer in the United States, or from Messrs Powell and Lealand in the United Kingdom, 1/16th, 1/10th and 1/20th inch objectives of numerical aperture of 1·5, at the present time there is great difficulty in finding lenses whose aperture is even 1·4 and none with the preceding high aperture is now made.

It seems that the current design tendencies in optical microscopy are working against the achievement of the ultimate in resolution; with modern equipment it is very difficult and perhaps even impossible to equal some of the feats of microscopic resolution which were reported half a century or so ago. The modern objectives, in addition to being of lower aperture, are often computed so as to deliver a truly flat field in order to make them more suitable for photomicrography and this may well mean the sacrifice of ultimate performance as far as resolution is concerned. With the development of the electron microscope, the stimulus to push the performance of the optical instrument to its limit seems to have decreased so much that it is not now commercially worthwhile for the manufacturers to provide equipment which could necessarily be of interest to relatively few workers and would in all probability be extremely expensive. The keen microscopist, still often an amateur worker as in the nineteenth century, who can obtain a second-hand Powell and Lealand stand No. 1 in good condition, and who can find one of their high aperture apochromatic objectives which is still in good working order is fortunate indeed. With such equipment, and a high degree of manipulative skill it is possible to obtain better results at the limits of resolution than with modern objectives, although for the majority of routine work in a laboratory the modern stand is much to be preferred on account of its ease of operation.

CHAPTER 6

The Optical Microscope in Modern Times

WE HAVE seen that by the end of the nineteenth century the optics of the microscope were excellent. The large microscope stands of the English makers were triumphs of the art of the instrument maker, and were much sought after by connoisseurs; they proved, however, to be too cumbersome and elaborate for the professional scientist, who preferred the lighter (and equally efficient) Continental instrument with its shorter tube length and Abbe illuminator in place of the elaborate and highly-corrected condensers of the English stand. Towards the end of the century, however, there was a tendency for English makers to move towards the Continental style of instrument; Powell and Lealand and Watsons, who were both producing excellent large stands, proved an exception to this trend. The large stands of Watsons had established an excellent reputation. One of their stands, that designed to the specifications of Henri van Heurck and bearing his name, is shown in Fig. 6.1 from which the main features of its construction are apparent. The instrument was furnished with two draw tubes, so that lenses corrected for either the English long tube or the Continental short tube could be used; there was a fine focus to the substage movement and the stage itself was provided not only with the full mechanical movements but also with rotating movements which are often very valuable in photomicrography. Such microscopes, and the rather simpler "Royal" and "Edinburgh Students" models of the same maker, although no longer in production may still be found in numerous amateur and professional laboratories in use and giving excellent service.

The period around the turn of the century proved to be a most important point in the development of the mechanical aspects of the

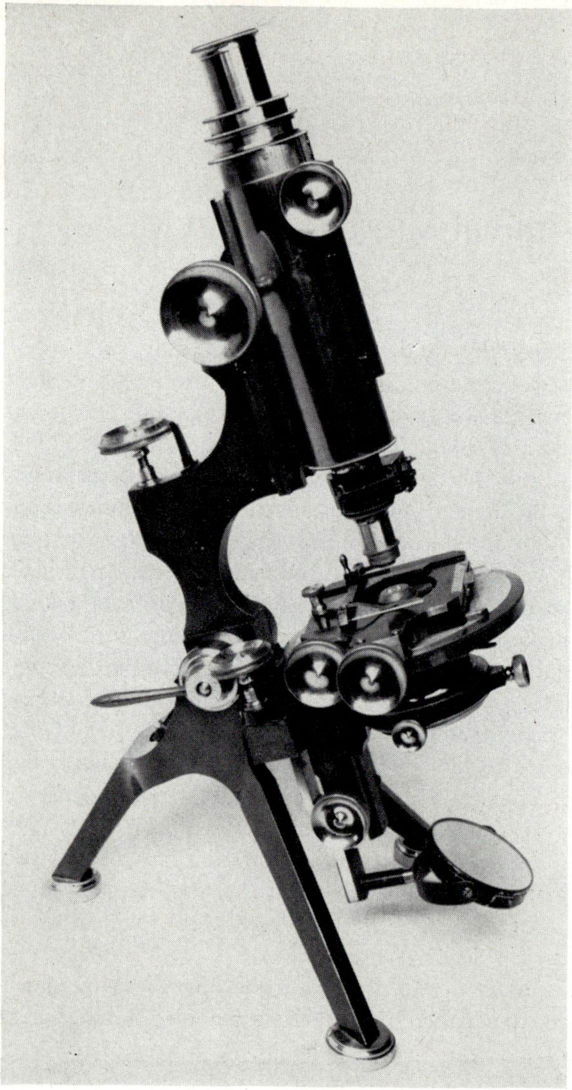

FIG. 6.1. The Watson "Van Heurck" microscope. Note the fine-focus adjustment for the condenser (located just behind the controls for the mechanical stage).

microscope. Before this date many microscope users had been the amateur workers who were concerned with observing pond life at low magnification or in seeking to prove by "diatom dotting" that the resolution of their latest high-aperture objective was better than that of the lens belonging to the next man. The twentieth century, however, resulted in a vast increase in the professional use of the microscope by medical students, scientists, and specialist workers in all fields of industry. Their needs proved to be quite different from those of the "diatom dotters"; they were usually content with lenses of a lower aperture, with a stand which was more manageable, and above all with apparatus which was sold at a reasonable price. It was this above all which led to the adoption of the Continental model, but one further change in the whole way of production of instruments also was involved. This was the change from hand construction methods to the machine-made, mass-produced article.

Formerly, microscopes were fashioned by hand, either by a single craftsman or by a very few workers, each specializing in his own particular aspect of the work. However, with the outbreak of the First World War in 1914 and the industrial demands then imposed, there were great advances in the methods for the mass-production of all types of instruments. Large numbers of gun-sights, telescopes, range finders and other instruments had to be made by relatively unskilled labour; at the same time high standards of accuracy had to be preserved. By the use of precision tools a great deal of the responsibility was removed from the operator. At the same time an advantage was obtained in that the microscopes or other instruments were all produced to standard tolerances and the individual parts were subjected to a careful inspection to ensure that they conformed to the requirements; this resulted in interchangeability of parts from one microscope to the next.

The development of standardization and mass production was a logical outcome of the moves leading to the interchangeability of component parts which had been initiated by the Council of the Royal Microscopical Society in 1857 when the advisability of having standard sizes for objective threads and for eyepieces and condenser mounts had been first suggested. The RMS standards and their modifications, which were added at a later date, were successfully instituted and eventually adopted by all the makers throughout the world. With the development of mass production the microscope could be purchased as a simple stand with the minimum of lenses and accessories, which could be added to as the occasion arose with the certain knowledge that all the extra parts would fit. A good example of a model of microscope which fulfilled the requirements of precision quantity production at a reasonable cost was the "Service" microscope of Watsons which was introduced just after the first war. To the basic stand such accessories as a centring substage, a rotating multiple nosepiece, binocular body and mechanical stage could be added any time as the need arose, so extending the scope of the instrument.

One notable tendency in microscope design during the last fifty years has been the increasing use of binocular tubes, which are now standard on almost every large research model. The use of a binocular tube allows both eyes to be employed simultaneously, which is important as it eliminates a considerable amount of the fatigue during the prolonged periods of observation which are often needed for research or diagnostic work. For low-power work there is the added advantage that some systems of adapting the microscope to binocular vision allow a stereoscopic effect to be obtained. This can be useful in understanding the morphology of certain small types of object such as the shapes of Foraminifera shells (Protozoa which are found in large numbers in the deposits of the ocean floor).

The binocular microscope has a long history, dating back to at least 1671 when Cherubin d'Orleans described such an instrument in his book *La Dioptrique Oculaire* and illustrated one six years later in his subsequent *La Vision Parfaite*. His microscope was in fact two separate microscopes, with completely independent optical systems, united with their tubes fixed at an angle so that they both focused on the same part of the object. No interest was shown in this system (which, together with its inventor, was criticized by Robert Hooke) until many years later when fresh interest was aroused by the publication of Wheatstone's paper on binocular vision in 1838. This started a search for the stereoscopic binocular microscope, which according to E. M. Nelson, was won by the Americans. Professor J. A. Ridell of New Orleans developed a workable system in 1854 which used two prisms placed behind the objective in the manner shown in Fig. 6.2a,

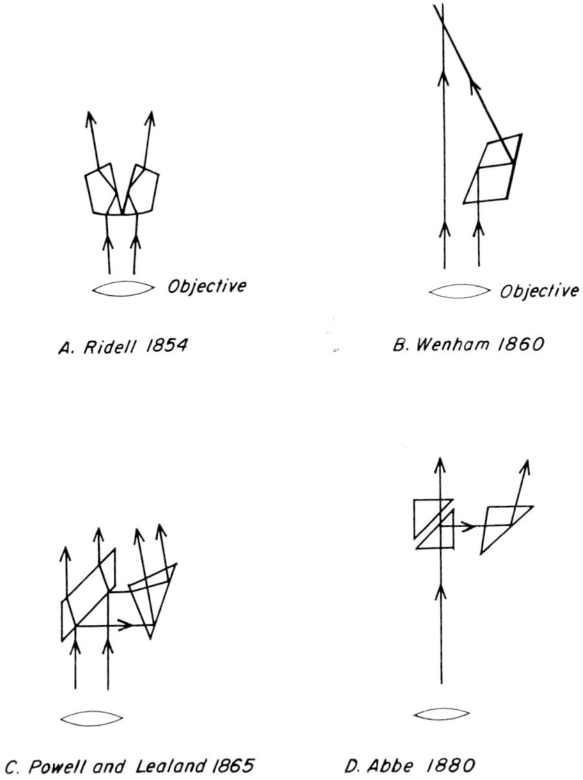

FIG. 6.2. The optical elements of four of the early binocular microscope systems.

so that the light from the objective was split and deflected up the two parallel eyepiece tubes.

Perhaps the most popular and successful of the nineteenth-century binocular systems was developed by Francis Wenham in 1860; his instrument employed convergent tubes with a beam-splitting prism inserted just above the objective lens to send light from half the aperture up the second tube to its own eyepiece. Wenham was associated with the firm of Ross, so that they manufactured many of these microscopes over the succeeding years. As no patent was taken out the system was soon extensively copied and most of the makers soon were offering binoculars constructed on this system. The arrangement of the prism (shown in diagrammatic form in Fig. 6.2b) is such that half of the aperture of the lens is occupied by the beam splitter which limits the usefulness of this system to objectives of low power. With these, however, a very beautiful stereoscopic effect is obtained. One advantage of this type of binocular is that the image in one of the eyepiece tubes is not altered at all or impaired in quality, whilst the effect on the other is only slight; the prism may easily be removed if desired, so converting the microscope back into a monocular with complete utilisation of the full objective aperture.

Powell and Lealand soon produced a modification of their own (see Fig. 6.2c) as also did Ross, which allowed the use of the binocular tube with the highest magnifications and with the full aperture of all lenses. With these systems, however, there was no stereoscopic effect and the only advantage was the lessening of fatigue. Many ingenious binocular systems, often involving numerous prisms and consequent light losses, were proposed and constructed in the latter half of the nineteenth century. Abbe himself designed a binocular eyepiece (Fig. 6.2d) which used two right-angled prisms, separated by a very thin film of air. The great drawback to this system, as to many of the other high-power binocular devices of this period was the unequal optical path in the two tubes which made some form of correction essential. Abbe provided dissimilar eyepieces to carry out this task. Another disadvantage of most of these early high-power systems was the unequal intensities of light in the two eyepieces; the Abbe binocular eyepiece suffered very much from this defect, as one eyepiece provided over two and a half times as much light as the other.

The first really successful high-power binocular system was introduced by Conrad Beck in 1913. It owed its inspiration to an invention by F. E. Ives in America in 1902 and used the so-called "Swan cube" as a beam-splitting device. This cube consists of two right-angled prisms cemented together after one of them had been coated with silver on the cemented face to such a thickness that approximately half the incident light was transmitted and half reflected. The Beck system is illustrated in Fig. 6.3a, from which it may be seen that this form of binocular also demanded the use of convergent eyepiece tubes; this Beck firmly believed to be the best system, as the eyes naturally converge for viewing near objects. In the same year Leitz produced their own system, designed by F. Jentzsche, which again used a beam-splitting prism embracing the whole aperture of the lens.

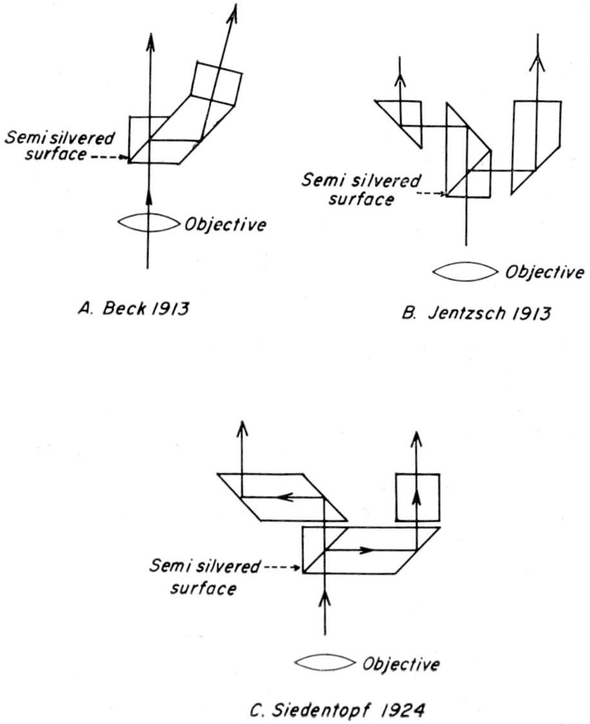

FIG. 6.3. Later binocular systems. Those due to Jentzch and Siedentopf are still in use.

In this form of binocular microscope the two eyepiece tubes are parallel, with the great advantage that a simple separation of the two outer prisms caused a corresponding variation in the interocular distance; the optical path difference could be very simply equalized by making the right-hand prism longer than the left-hand one (Fig. 6.3b). This system was so successful that it is still used today as the basis of most of the binocular microscopes now produced. Another variant which found some favour is due to Siedentopf and was invented around 1924; this is shown in Fig. 6.3c, and it will be seen that the interocular distance can be varied by a simple rotation of the left-hand prism. This has no effect on the optical path difference and so with this system variation of the interocular distance has no effect on the tube length of the microscope.

Another twentieth-century trend has been the adoption of more or less complicated prism systems so that the eyepieces may be inclined whilst the limb is vertical and the stage of the microscope remains horizontal. Such devices provide a very convenient viewing position for the operator allowing the instrument to be used on a table whilst the operator remains seated normally. Two microscopes, one by Zeiss the other by Wild, are shown in Fig. 6.4. From this illustration the appearance of a modern research biological microscope may be gauged. The inclined eyepiece tubes are clearly visible and the general solidity of the main limb casting will be noticed. This is a very important trend in modern microscope design and the solidity of such stands (preventing excessive flexure of the limb during use, which may be troublesome with high-power lenses) is a great advance over the earlier instruments. Indeed, in some of the more elaborate microscopes this is carried so far that the weight of the casting makes it very difficult to lift the microscope from the table! The contrast with the microscope of fifty years ago, as represented by the van Heurck instrument shown in Fig. 6.1, is very marked.

In the current research microscope, as for example in those illustrated in Fig. 6.4, it can be seen that the focus controls, both coarse and fine, have been made concentric so that they are very convenient to use and they have been brought down to a low level; here we have in effect a reversion to the idea of George Adams the Elder, where he made the focusing arrangement of his "New Universal" microscope of 1746 at table level (see Chapter 4, Fig. 4.6). In some modern instruments, however, the body tube and the lenses do not move in order to focus the object; in the interests of stability the limb and the lenses have been fixed and the focus is varied by the movement of the stage. It is claimed that this allows a much steadier focusing motion and facilitates the use of very high-aperture lenses.

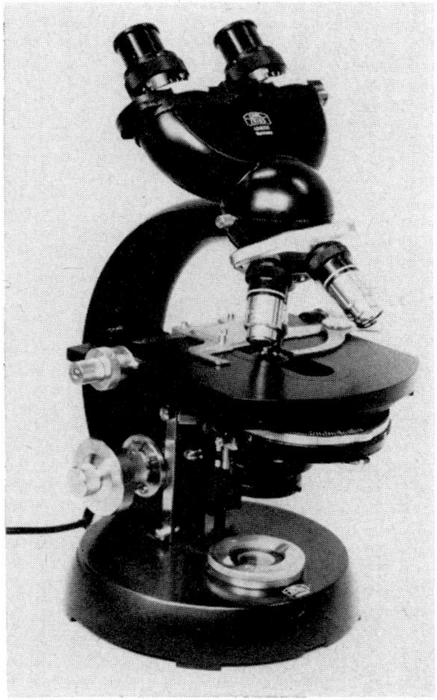

(a) By Zeiss.

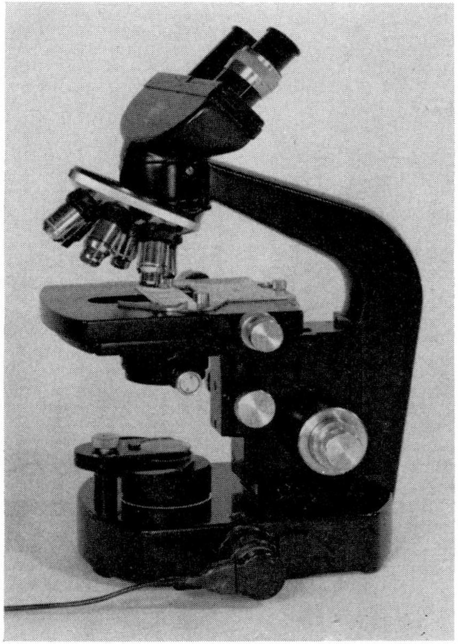

(b) By Wild.

FIG. 6.4. Two modern Continental research microscopes.
Both of these microscope stands possess built-in illumination; the controls of both instruments are now placed below the stage level.

One further feature of modern practice which is universal in research quality stands and is rapidly spreading to simpler microscopes of the student category, is the incorporation of the lighting unit into the base of the microscope. The low-voltage bulb is housed in the base of the instrument and its light concentrated by a lens system and reflected onto the condenser by means of a right-angled prism. This has the advantage that the whole microscope system is rendered much more portable and it is easier to set up and align for use. The low-voltage bulbs used in these lamps have a filament composed of several coiled strands of wire arranged side by side in order to provide a large or "solid" source. It is thus easy to form an image of the source that fills the full aperture of the condenser. The low-voltage lighting unit gives a very intense light which is suitable for observation at the highest powers and also for photomicrography. For normal visual observations the light is dimmed by reducing the voltage by means of a suitable rheostat or variable transformer.

Such modern microscopes are provided with relatively few accessories. If one looks at a large research microscope of even sixty years ago one is struck by the large number of accessories which often necessitated a second box as large as that of the microscope itself; most of these accessories were used for varying the type of illumination, and with the better appreciation of the principles underlying the microscope theory they naturally were all rendered obsolete. Although the modern instrument comes with few accessories it is so well engineered that it proves an easy matter at any time to replace certain parts such as the body, the stage or the condenser with components of a different type which may be required for some particular purpose. Some idea of the various range of fittings available may be gained from Fig. 6.5 which shows the basic microscope stand and some of the variant arrangements which may be bought and set up at any time for the accomplishment of special tasks.

The whole design of the modern microscope stand is extremely functional, all its parts being arranged for ease of manipulation. In this respect our present microscopes show a marked advance over the older stands; although the best of the old microscopes would do the job just as well as a modern microscope, the latter is much easier to use and adjust. Some flexibility, however, has had to be sacrificed in order

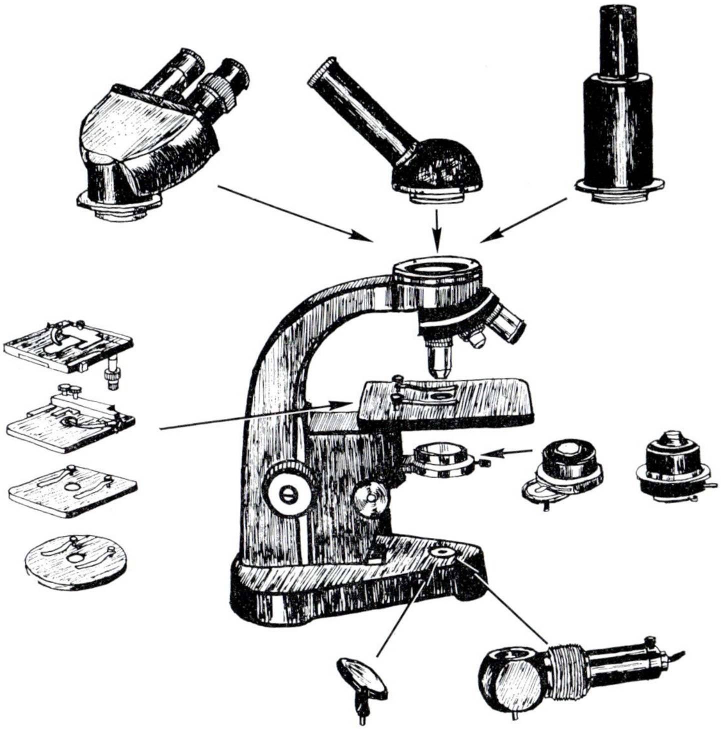

FIG. 6.5. Diagram illustrating some of the various components which can be added to a modern basic microscope stand.

to achieve this ease of operation and it is not now the practice to make each individual microscope stand universal in function. If polarized light is required, or if phase-contrast or some other special technique is needed, then it is more usual nowadays to employ an instrument specially designed for the job rather than proceed to adapt a standard biological microscope by means of added accessories.

In its basic design, the microscope has changed remarkably little since the nineteenth century, each improvement tending to be in detailed construction rather than representing a radical rethinking of the whole design. Recently, however, one of our leading manufacturers (Vickers Instruments) has produced an instrument which represents a breakaway from traditional microscope design. This microscope, known as the "Patholux", is shown in Fig. 6.6, from which it can be seen that the traditional limb is replaced by a very solid casting, which has considerable rigidity and carries the stage and the optical system.

It can be seen from the figure that the limb now takes the form of a box frame, perhaps one of the most stable of all engineering constructions; the stage is very large and rigid and is firmly fixed, the focusing movement being once again applied to the lens system. The operating controls are arranged on the base panel and it has been found by ergonomic study that this placing allows the maximum of comfort allied to operating efficiency. The coarse and fine focus controls are concentric and it is interesting to note that in this microscope, for perhaps the first time for a hundred years, the rack and pinion mechanism for the operation of the focusing has been abandoned; in its place a sturdy lever and scroll mechanism has been substituted. There is an integral light source and the whole instrument has been designed to be as adaptable as is consistent with the maintenance of the primary functions of the microscope stand. This microscope may be compared with the instrument shown in Fig. 6.7 which represents the conventionally designed instrument from the same maker, with the focusing movement of the instrument being applied to the stage and with the standard type of base and limb.

With the rethinking of the engineering of the microscope stand, changes, although less striking, have also been made in the optical system. For a long time after the introduction of apochromatic lenses it was thought that the ultimate had been reached; this, as so often in the past, proved to be a mistaken view. Improvements have taken place in the methods of mounting the objectives on the instrument, so that the modern microscope now carries three, four or even six lenses so arranged that any one may be instantly brought into use. This change of objectives by simple rotation of a nosepiece is not, of course, a new development; it was used in the eighteenth century by Benjamin Martin among others. The modern revolving nosepiece, however, is a piece of high precision engineering so that the centration of each lens to the optical axis of the microscope is guaranteed.

One important feature of the modern objective which has been developed along with the revolving nosepiece is their adjustment so

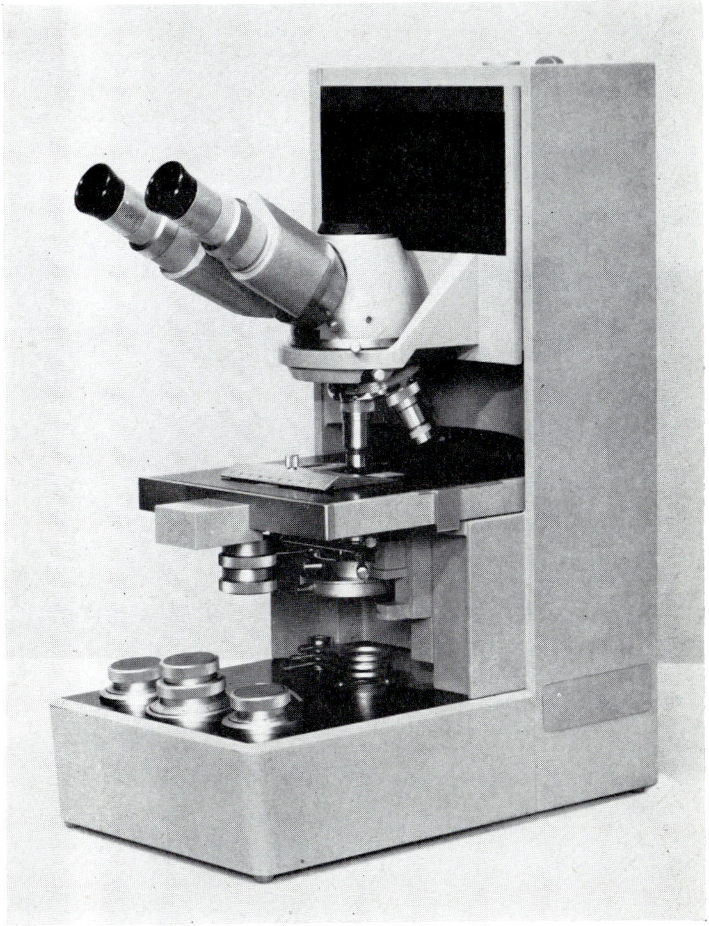

FIG. 6.6. A modern trend in microscope design. The Vickers "Patholux" microscope.

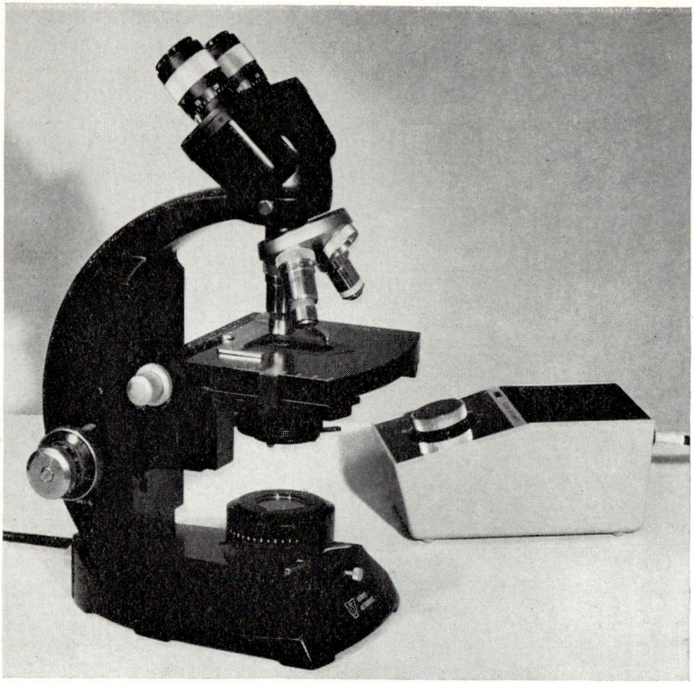

FIG. 6.7. The Vickers research microscope model "M15b". The light control unit is seen in the background.

that all the objectives on one nosepiece are approximately in focus when this is rotated. This is called "parfocality" and is very valuable as in routine diagnostic work it is very often necessary to inspect an object with several different powers in rapid succession; much time and temper (not to mention accidents to the objectives!) are saved if the nosepiece can be rapidly rotated and the new lens will still be approximately in focus.

One danger which is always present, especially with the higher powers in which the lenses have a very short focal length and consequent short working distance, is that the objectives may inadvertently be racked down into the slide during the focusing movement. In the past many valuable lenses and preparations have been ruined by this;

even a very slight knock is often sufficient to displace the front element of some oil-immersion lenses. In recent years a very simple device has been adopted to prevent damage in this way. The lenses of the objective are arranged so that they can slide in a cylindrical fitting in the upper part of the mount, against the pressure exerted by a weak spiral spring. In normal use this spring keeps the lens fully extended in the correct working position, but if the front lens is accidentally brought into contact with the slide then the full pressure exerted by the focusing screw is absorbed by compression of the spring, the lens elements of the objective sliding into the mount; the result is that no damage to the mounting of the lenses can take place with normal usage.

In addition to these improvements in the mounting of the objective, there are some other improvements which are worth noting. First, there is the much more extensive use of the mineral fluorite. The use of natural fluorite by Abbe and other makers has already been mentioned, but pieces of natural fluorite large enough for optical use are now very rare indeed. This placed a very severe restriction on the use of this substance for the components of microscope lenses. In recent years, however, it has been discovered that the mineral can be recrystallized and obtained as discs which may be several inches in diameter and without flaws and so perfectly suited to optical work.

This "synthetic fluorite" (and the use of crystals of lithium fluoride which may be synthesized in the laboratory) has enabled this substance to be used extensively in the construction of microscope lenses; by its aid the designer has been able to obtain superior corrections at a very reasonable price. The "fluorite" lenses have corrections intermediate in quality between the achromatic objective and the very expensive apochromat and are usually referred to as "semi-apochromatic" objectives. They possess the improved spherical correction associated with apochromats and are especially valuable for photomicrography. Both fluorite and apochromatic objectives, despite their very high degree of spherical and chromatic correction, do suffer from a rather marked curvature of their field. This means that the only area in sharp focus at any one setting of the fine focus control is very limited. If the centre of the field is sharp, then the edges are out of focus and vice versa. For visual examinations, especially where one is working at high resolution, this is not a great drawback, as it is natural to bring the object of interest to the centre of the field for examination; the expert microscopist is also constantly manipulating the fine focus, so that field curvature is often passed almost without notice.

With the development of photomicrography as a means of recording the microscopic image this defect has assumed a much greater importance. Only one plane of focus can be recorded at any one time with the camera and the out-of-focus areas around the periphery of the field then become very objectionable. To some extent this fault may be overcome by using sufficient camera extension (or a high-power eyepiece) so that only the sharp central portion of the image is recorded on the plate; this solution may, however, involve the use of excessive magnification so that the resultant picture suffers from the effects of "empty" magnification in which blurring ensues and no further detail is rendered visible. The solution to this problem has been provided by the optical designers who have computed lenses which not only have very good corrections but also provide a truly flat field. This has been achieved only at the expense of making a very complicated objective and at the expense of a certain amount of resolution. These "plan-apochromats", as they are called, may now be obtained with various focal lengths, but as they combine the corrections of an apochromat with the additional quality of a flat field, naturally their cost is extremely high.

One of the great aims in microscope design over the last hundred years has been a desire to introduce compactness into the design so the instrument could be made portable for use in the field. We have seen that the "chest" microscope of Nairne (p. 93) was designed with this end in view. Most of the leading makers in the nineteenth century also included some form of small microscope, often made to fold up for ease of transport, in their catalogues. Most of these stands, however, ingenious though they were, represented no new departure in design but were merely a simplification of the basic large stand with the elimination of non-essential components and the provision of various devices for packing the whole into a confined space. In 1934 a new microscope was announced which represented a fresh approach to the whole problem of design of a portable microscope. This instrument was produced by McArthur especially for diagnostic field work. He stipulated that any portable microscope must be light, occupy very little space and be capable of operation when held in the hand. In addition such a microscope must be easy to set up and be capable of producing work of exactly the same quality that a larger laboratory microscope could achieve.

McArthur pointed out that the majority of the previous attempts to produce a portable microscope had merely been modifications of the conventional instrument and had resulted in the production solely of a *collapsible* microscope, which was inevitably a compromise between lightness and rigidity with all the subsequent inconveniences of setting up. In consequence, he produced a radical departure from the standard design by using a standard objective which was set to look upwards at an inverted slide. The eyepiece was mounted in the normal way but between the objective and the eyepiece were interposed two reflecting surfaces so that there was no conventional tube; instead the body was machined from a single casting. This is shown clearly in Fig. 6.8 from which the size of the instrument may be gauged.

The separate objectives were mounted on sliders which not only enabled a very quick and precise change from one power to another but also provided very accurate centring of the lenses. This slider was made with an L-shape and was fitted with provision for upwards and downwards movement in a dovetail; this was actuated by a screw and lever mechanism in order to provide the means of focusing the lenses. There was no coarse adjustment on this instrument, as with an inverted slide and parfocal objectives the need for this device is no longer felt. In actual practice it is found that the fine focus only needs a very slight amount of adjustment between the different lenses. McArthur claimed that the use of the inverted specimen slide also led to other advantages, such as a more readily accessible condenser and easy location of objects on the slide. As the body was made in a single casting the whole arrangement showed a very great rigidity and the objects of portability and ease of operation were secured.

Since its first inception this microscope has been considerably developed and now it forms a complete system of microscopy, with provision for fitting built-in illumination, photomicrographic attachments and a whole host of other accessories. Such is the rigidity of the construction that actual instruments have been run over by a motor car and dropped from an aeroplane without in any way impairing their

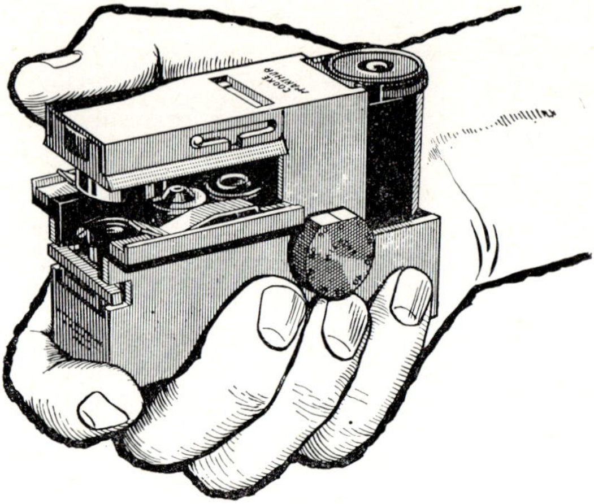

(a) The McArthur portable microscope.

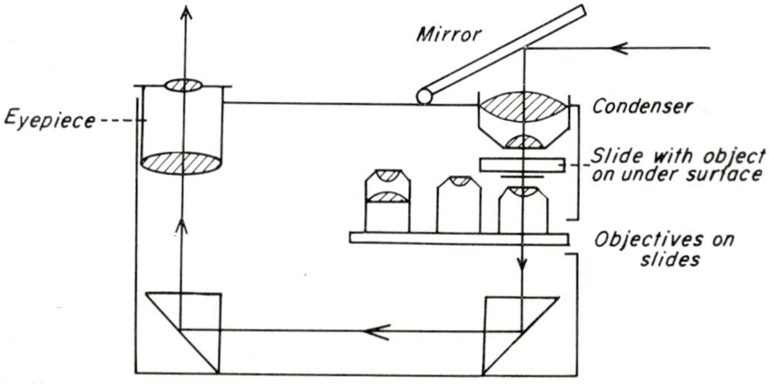

(b) A simplified section of this instrument to show the three objective lenses on a slider, together with the prisms and the eyepiece.

FIG. 6.8.

performance. Although primarily designed for diagnostic field work, for example, in the study of malaria in the tropics, it is now proving to have a whole new field of applications especially in some industrial concerns where by means of the incident light attachments it may be placed directly onto a metal or other surface in order to examine the surface structure.

With the development of the apochromatic objective it seemed that the optical advances possible in the light microscope had approached the limit; indeed apart from the developments listed above, which did not affect the resolving power of the instrument, little real progress in the optics of the microscope was evident for several decades. In 1934, however, a paper was published by F. Zernike on a new method for testing telescope mirror systems. In the following year the same worker extended this to microscope objectives and he initiated the method of "phase-contrast" microscopy which has revolutionized biological microscopy by allowing the direct study of transparent objects.

With a microscopic object not only must the detail in the image be resolved but also it must be rendered visible; this is normally achieved by changes in the intensity of the light waves which pass through the object (see Fig. 6.9). Such changes in the amplitude of the light waves are produced by the absorption of some of the light in the specimen, a phenomenon which is greatly enhanced if the latter is coloured either naturally or by means of staining procedures. If staining is not possible for any reason, then visibility may often be achieved by modifying the illumination system, either by the use of very oblique light or by dark-ground techniques in which the light is so oblique that no direct light falls within the acceptance angle of the objective and only the light diffracted from the object contributes to the formation of the image.

Biological specimens, such as living cells, are normally very transparent and often it is not possible to stain them without causing their death and possibly altering their structure. Such objects form a large part of the interest in biological microscopy and therefore until some method of adding contrast to their image could be developed the microscope could not realize its full potentialities in terms of resolution on living biological material.

Fortunately, however, such biological material does usually exhibit differences in refractive index between its various component parts and the mounting medium. Although the cells do not change the amplitude of the light passing through them, they do affect the light waves. Because of these differences in refractive index, there is a difference in the optical path of light which passes through the specimen compared to that which passes around it. In consequence the light waves which pass through the object and its component parts will be changed in phase (Fig. 6.9). Our eyes are completely incapable of

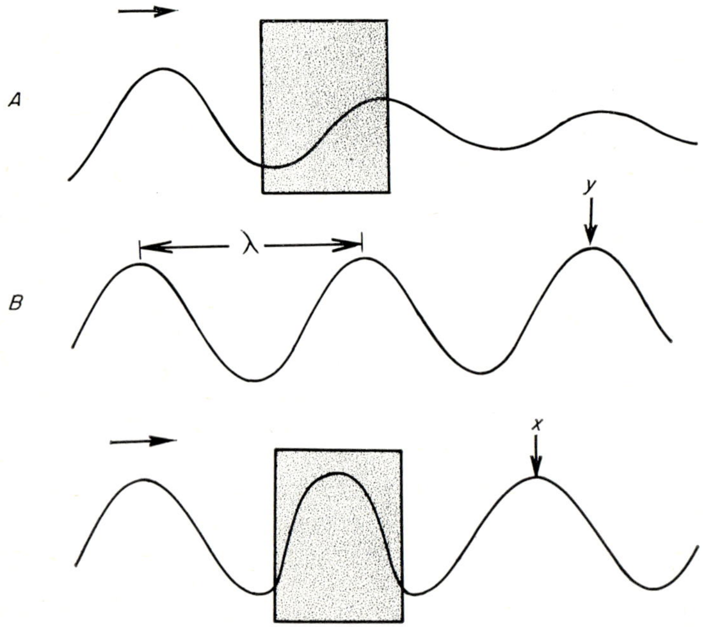

FIG. 6.9. A diagram to show the effect of two types of object on the light waves passing through them.

(a) An absorbing object causes a change in amplitude of the wave only.

(b) A transparent object causes no change in the amplitude of the waves, but alters their phase with respect to the original beam.

Note that the crest of the wave x is retarded relative to that of wave y.

perceiving such changes in phase of light and so this type of specimen is, therefore, normally invisible. Zernike's method was the first practical way of making such phase differences into observable differences in amplitude and so by optical means making transparent objects visible through a microscope.

With biological objects the difference in optical path introduced by the object is small. It depends on the thickness of the object and upon the difference in refractive index between it and the medium, but in most cases the phase difference between the direct light (the zero-order beam) and the first-order diffracted light is about one quarter of a wavelength; this is usually expressed in the form $\lambda/4$. Zernike's simple but ingenious idea was to introduce a further $\lambda/4$ path difference between the direct and diffracted light so that the total difference of phase between the two beams was increased to half a wavelength; at this value destructive interference between the wave fronts would occur, so causing the desired differences in amplitude. As Zernike wrote in 1942:

> By the phase-contrast method transparent details of the object which differ in thickness or in refractive index, appear as differences of intensity in the image.

The desired additional retardation between the diffracted light and the zero-order beam was introduced by the inclusion of a "phase strip" into the optical system. Zernike at first used straight phase strips but he also studied other shapes such as the cross and the annulus or ring form. A diaphragm of a similar shape to the phase strip in use is placed in the back focal plane of the condenser so that it will, by the principle of the conjugate focal planes of the microscope, be imaged directly in the back focal plane of the objective. It is at this point that the phase strip or phase plate, as it is called today, is mounted. The phase plate consists of a plane parallel piece of optical glass which has an area corresponding to that covered by the image of the diaphragm either raised or recessed. The depth of the recessing is such that the difference in optical path through the recessed area is less by a quarter of a wavelength than that through the remainder of the glass phase plate. Such a system is shown diagrammatically in Fig. 6.10. It will be obvious that all the direct light from the annular diaphragm (D) will pass through the condenser and the objective and be imaged on the recess in the phase plate (P) which is in the back focal plane of the objective.

The phase plate, with its recess, is shown in cross-section in the lower part of the diagram (Fig. 6.10) and the direct light (DL) is shown passing through the area of the annular recess. The diffracted light (which it

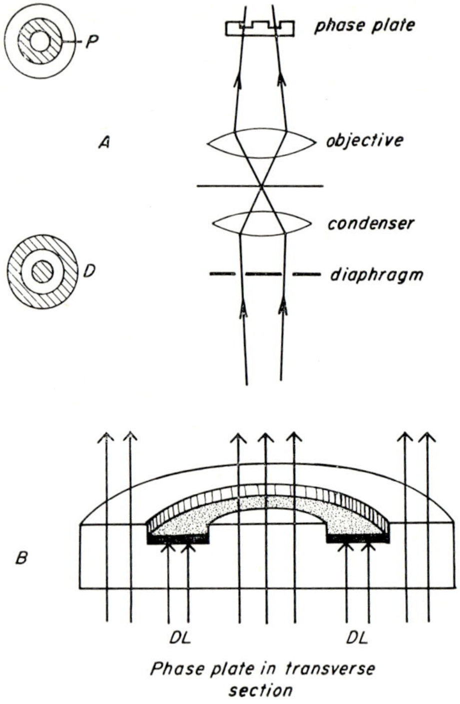

FIG. 6.10.

(a) The optical arrangement of a phase-contrast microscope. The annular diaphragm (D) and the phase plate (P) are shown on the left.

(b) An enlarged section through a typical phase plate. The direct light (DL) passes only through the groove in the plate whilst the diffracted light traverses the whole structure. An absorbing layer is deposited in the groove of the phase plate to reduce the intensity of the direct light with respect to that of the diffracted rays.

will be remembered originates at the object and which is, therefore, retarded by $\lambda/4$ with respect to the direct light) will not be restricted to this area of the phase plate as it is not conjugate with the focal planes of the condenser or objective and will, therefore, pass through the whole area of the phase plate. This is a much larger area, indicated by the unlabelled arrows in the diagram, and in so doing it will have to pass through a greater thickness of glass which will introduce a corresponding extra degree of retardation. As this thickness is so calculated to give the extra $1/4\lambda$ which is required to make the direct and diffracted light interfere destructively, the necessary contrast will be introduced into the image. Zernike further introduced a layer of absorbing material into the phase strip which served to reduce slightly the intensity of the direct light and so still further improve the contrast. This description of the principle of the phase-contrast microscope is, of course, extremely simplified and interested readers are referred to the standard books on microscopy for a fuller explanation.

Zernike obviously realized the tremendous implications of his method very early on, perhaps as early as 1932, before the original paper was even published. Patent rights were secured in collaboration with the firm of Carl Zeiss, who proceeded to experiment on practical methods of producing and exploiting the invention. A phase-contrast microscope was in fact produced by Zeiss in the early years of the Second World War, but such instruments did not become generally available until after the cessation of hostilities.

In this country, C. R. Burch was engaged in research on this new system of microscopy and in 1942 he published a paper in collaboration with J. P. P. Stock on experiments they had carried out with straight strip phase plates. This paper gave full details of the methods for constructing experimental phase plates and, equally important, it included some results showing the application of this method to the study of unstained living leucocytes and their interactions with bacteria in pus.

On the Continent other workers were using phase-contrast systems for the study of living cells; one notable piece of work was the use of this new method by K. Michel to obtain by time-lapse techniques a remarkable cine-film of the divisions and changes which occurred in the development of the sperm cells of the grasshopper. When this film became widely known to the biologists it, more than any articles or demonstrations, convinced them of the tremendous potentialities of the phase-contrast method for the study of the activities of living cells. Although early work was carried out with slit and cross shaped phase plates, it very soon became apparent that the annular shape of phase plate possessed considerable advantages as this shape had a complete axial symmetry and it did not introduce any undesirable asymmetry into the image as did the other forms of plate.

After the end of the war microscopes based on the phase-contrast principle became available commercially and research both into the applications of the system and into its theory proceeded very rapidly. It was realized that phase-contrast does introduce a slight loss of resolution but this is far outweighed by the visibility of detail which it confers on unstained or living material.

The phase-contrast microscope has the tremendous advantage of simplicity and it has now become established as a routine laboratory instrument. It has fully justified its early promise and many valuable discoveries have been reported by its aid. The image is not perfect, however, the edge of the object being surrounded by a very prominent halo, for example. This is due to shortcomings in the optical system whereby the separation of the direct and diffracted light is effected; some of the diffracted light does in fact pass through the area of the phase plate which is intended to receive the direct light and it is this mingling which gives rise to undesirable optical effects.

To a large extent these defects have been eliminated in a further development which is known as the interference microscope. As we have seen the contrast arises in the phase microscope by interference between the direct light and the diffracted light which arises at the object and which has undergone a retardation in phase. In the interference microscope the destructive interference takes place between two wave fronts which are produced within the microscope optical system itself. One set of waves is allowed to pass through the object and is subjected to modification by it, whilst the other wave front (which is referred to as the reference beam) passes along a different course which does *not* include the object.

One of the obvious solutions was to use a double beam system with in effect two duplicate microscopes with matched condenser and objective systems. One of these carried the object whilst the other had a blank reference slide. This idealized system has been called the "round the square" microscope and is shown in Fig. 6.11a.

In this microscope a beam of light is divided by means of semi-silvered reflecting surfaces and passes through the two microscopes. The specimen introduces a phase change into the light which passes through it and after the recombination with the reference beam which has traversed the other arm of the square the desired interference takes place. Such a microscope was not considered feasible for many years owing to the expense of duplication of the optical systems and the difficulties in alignment of the system, but recently the firm of Leitz has developed a form of interference microscope working on this principle. The ray diagram of their instrument is shown in Fig. 6.11b from which it may be seen how closely this system approximates to the ideal "round the square" interference microscope. Such a system with its complete physical separation of the two beams by means of beam-splitters using semi-reflecting surfaces has many advantages and for some purposes works rather better than the other current instruments in which the separation of the beams is effected by some form of birefringent crystal.

In 1947, F. H. Smith developed what proved to be the first really practicable interference microscope which was efficient at all powers. This microscope used birefringement elements to separate and recombine the two beams. It is not possible to go into a detailed optical description of Smith's microscope here; such information is available in the technical literature. It will suffice to mention that Smith produced two variants of the system, one involving a lateral displacement of the reference beam with respect to the object, the so-called "shearing" system, and the other involving a "double-focus" arrangement whereby the reference beam is brought to a focus below the level of the object and so the latter is surrounded by an annular cone of the reference light which does not actually traverse it. This will be clear from the diagrams of the two systems which are provided in Fig. 6.12.

This interference microscope possesses several advantages over the simpler phase-contrast instrument, in that, for example, the image is not surrounded by a troublesome halo. Also in the phase microscope

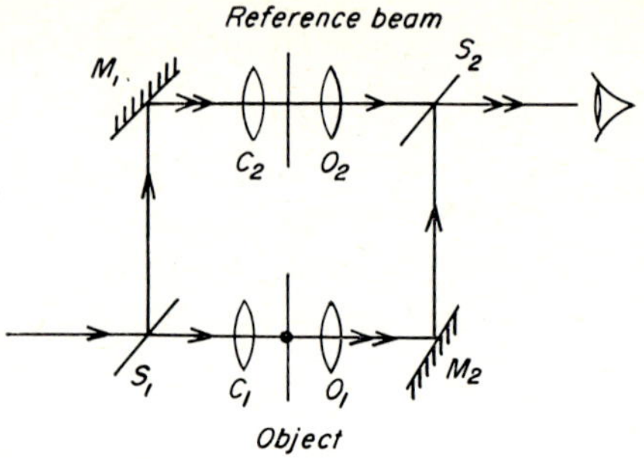

(a) An idealized "round the square" interference microscope system. S1 and S2 are semi-silvered surfaces which reflect and transmit equal amounts of the light falling on them. C1 and C2, and O_1 and O_2 are matched pairs of condensers and objective lenses respectively; M_1 and M_2 are totally reflecting surfaces.

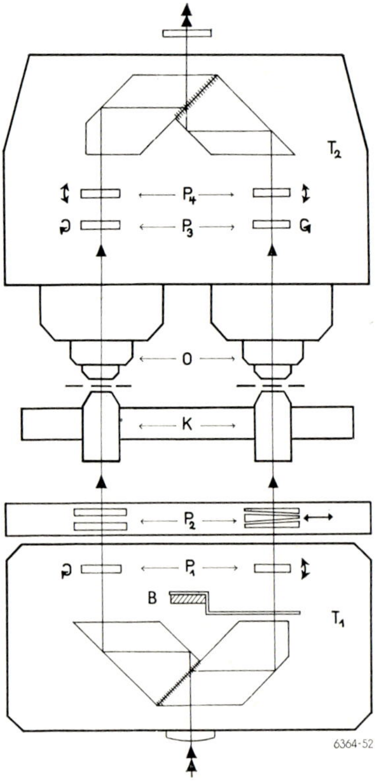

(b) The optical construction of the Leitz interference microscope which utilizes the same principle.

FIG. 6.11.

the contrast is only adequate in those regions where there is an abrupt change in optical path difference, such as at the edge of an object; in the interference microscope on the other hand gradual changes in optical path difference are equally well shown. This makes the interference microscope a particularly valuable tool for morphological

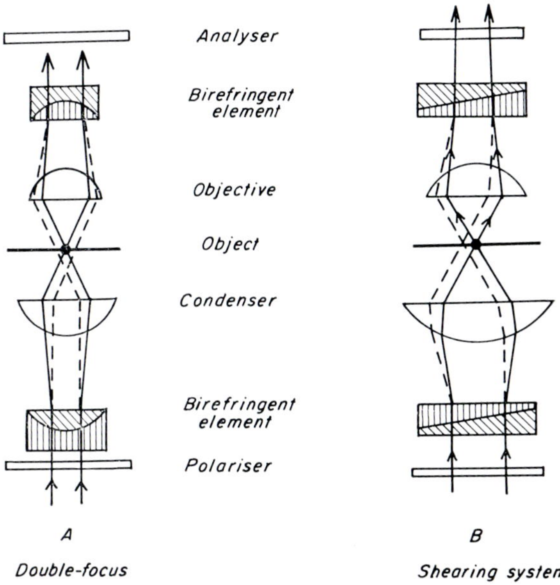

FIG. 6.12. The optical principles of the Smith interference microscope system using birefringent material to effect beam separations.
(a) The double-focus variant, in which the reference beam is brought to a focus at a different level from that of the object beam.
(b) The "shearing" type, in which the reference beam is displaced laterally with respect to the object.

studies of cell inclusions which are small (and therefore show little phase change) or which have a refractive index very close to that of the surrounding cytoplasm. If the instrument is used with white light then the interference often removes only a portion of the wavelengths which are present so that the image appears in the subtractive colours of Newton's scale. This again is a feature of great advantage for morphological work, as the eye is very sensitive to changes in colour whereas small changes in greyness as might be produced with the phase-contrast microscope would probably pass unnoticed.

The most important advantage of the interference microscope, however, is probably the fact that the optical path difference introduced by the object may be measured directly and very easily. This measurement allows a calculation of the refractive index of the inclusion and also determination of its dry mass. Recent developments, especially in America, indicate that in the near future electronic methods of modulating and changing the phase relationships of the light will be used in microscopy. Although these systems suffer from the disadvantage that they are complex and still highly experimental, they increase the accuracy with which phase change measurements can be made and they speed up the process of obtaining quantitative data. Such information is valuable as it is possible by the use of phase-contrast and interference microscopy coupled with time-lapse cinematography, to build up a picture of the changes which take place throughout all the stages of, say, a single division cycle of a cell in tissue culture.

With the development of such techniques which enable contrast to be added to the image of cells or living biological preparations by optical means, whole new fields of study are opened to the biologist; it is by the use of new methods, as those which have been mentioned above, that the optical microscope is contributing very greatly to scientific research at the present time. The emphasis has changed completely from that which prevailed only a few years ago, and much less interest is shown in trying to work at the ultimate limit of resolution. From the theoretical studies of Abbe it was clear that the resolution of the microscope (R) was given by the general relationship

$$R = \frac{0 \cdot 61\lambda}{\text{N.A.}}$$

where λ was the wavelength of the light used to illuminate the system and N.A. referred to the numerical aperture of the objective. For various practical reasons the maximum possible upper limit of the numerical aperture of the objective had been attained some time in the latter years of the nineteenth century; hence if the system were to give a greater resolution then it must be achieved by using light of shorter wavelength. This was, of course, attempted many times and slight improvements in resolution were obtained by skilled microscopists such as Woodward and van Heurck who habitually used deep blue light for their photomicrographs.

These tendencies were carried to their logical conclusion in 1904, when Köhler and von Rohr of Zeiss designed a microscope intended to operate in the ultra-violet region of the spectrum. This instrument used the radiation generated by a spark (actually a cadmium arc was used) and it was based on the standard vertical microscope which this firm was manufacturing at that time. According to Abbe's theory, a microscope operating with radiation of a wavelength of 2750 Å would be expected to show double the resolution which it would afford with light of wavelength 5500 Å and objectives of the same numerical aperture. In effect, a lens of N.A. 1·25 specifically designed for operation in ultra-violet would be equivalent in resolving power to an objective with a working aperture of 2·50 operated in green light at 5500 Å.

Although this twofold increase of resolution obtained by the use of ultra-violet radiation is very well in theory, there are some practical drawbacks to the regular use of such short wavelengths in practice. Not least is the fact that the eye is not sensitive to radiation in this region of the electro-magnetic spectrum. This means that all focusing must be carried out with an artificial aid such as an eyepiece fitted with a fluorescent screen to convert the short wavelength radiation into visible light. The final image has to be recorded on a photographic plate and all the observations made on the final print. It is, of course, possible to focus the instrument in visible light and then change to illumination with ultra-violet but if this is done then the focus of the object has to be altered by a predetermined amount in order to compensate for the fact that the focus of the ultra-violet rays is not equivalent to that for visible light. This disadvantage, inherent in all lens systems, may, however, be avoided if a mirror objective is used; this is one field in which the reflecting microscope has a valuable role to play.

A further disadvantage of using short wave-length ultra-violet radiation in the microscope for increasing the resolving power is that the glass of which the lenses are made will only transmit such radiation freely down to a wavelength of about 3000 Å; below this figure glass is not suitable as a material for lenses. This difficulty may be avoided by making the lenses out of synthetic fluorite or even better, of fused quartz. This latter substance was in fact used by Köhler and von Rohr in their original microscope and it is still favoured today for lenses which are intended for critical work in the far ultra-violet. Not only are the various optical glasses opaque to the shorter wavelengths of ultra-violet, but so are also the normal mounting media and immersion oils, and hence substitutes have to be found. It proved possible to mount the specimens in water or castor oil or in glycerine jelly, whilst glycerine was found to be suitable for the immersion medium between the coverslip and the front element of the lens. Again, the coverslip and the actual slide on which the specimen is mounted must be made from fused quartz, as also must the substage condenser, so it is apparent that if serious work is intended in the far ultra-violet it is a matter of some complexity, not to say expense!

At the present time ultra-violet microscopy is no longer in any demand as a means of increasing the ultimate resolution of the optical microscope. Other valuable information has been derived from its use, however; studies initiated in the late 1930's by Caspersson, and since continued by him and his pupils, utilized the heavy absorption of certain wavelengths of ultra-violet which is characteristic of compounds such as nucleic acids, to locate by photometric means their distribution within the cell, and to follow changes in them during certain activities such as division of the cell.

Today, ultra-violet (or deep blue light) is more commonly used in what is termed "fluorescence microscopy". Here the short wavelength radiation is used as a source of energy to stimulate substances present in the tissue on the slide. These compounds (which may be already present or which may be added during the preparation of the specimen) absorb the energy and re-emit it at a longer wavelength, i.e. in the visible spectrum, in the form of fluorescence. For this type of microscopy the requirements are much less stringent; usually the source is a mercury arc lamp provided with suitable glass filters to absorb the visible light and isolate the desired band of shorter wavelengths.

As the radiation is only acting as an "exciting" agent it is possible to use the longer wavelengths and so the optics of the microscope may be made out of glass. It is necessary to ensure that all the ultra-violet light which is not absorbed in stimulating fluorescence in the section is prevented from passing up the microscope and entering the eye, so that such instruments invariably contain a selective ultra-violet absorbing filter in the optical system somewhere between the objective and the eyepiece. This protects the eye from the damaging effect of ultra-violet radiation and does not detract from the efficiency of the instrument since the important information is contained in the fluorescence, which is, of course, light in the visible region of the spectrum.

The fluorescence microscope has assumed a great importance in medical research today, for by its aid it has proved possible to locate the sites of antibody production in the body. This is done by coupling fluorescent dyes (which serve as labels) to the compounds which, when applied to cells or tissue sections, link in their turn to the proteins known as antibodies which form one of the body's main defence mechanisms against disease. Such specialist uses of the fluorescence microscope are, however, beyond the scope of this book; they form a striking development from the system which was developed by Köhler and von Rohr in the early years of the century in a simple quest for more resolution.

Up to the middle of the present century the optical microscope was regarded entirely as a tool for studying the fine details of the morphology of an object, which in the case of a cell meant a preparation which had been killed and prepared according to an elaborate schedule and stained with the synthetic dyes. With the coming of the electron microscope the optical study of structure has receded in importance and the optical microscope is being used for important studies of the living cell and organism; more and more the microscope is being used as an analytical tool to provide information about the chemical and physical components of the cell. One of the most striking developments in recent years is the introduction of automation to microscopy. This is apparent firstly in the simple photometric devices which are used to measure the intensity of light passing up the microscope in order to estimate correctly the duration of a photographic exposure. More recently sophisticated automatic equipment has been manufactured which not only measures the intensity of the light but also controls the opening and closing of the photographic shutter when a predetermined amount of exposure has been given to the film.

For 300 years the human eye and brain have been used in conjunction with the lenses of the microscope. When the emphasis was upon the acquisition of purely morphological data, this proved to be a perfectly satisfactory combination; now, however, with the change in emphasis and the desire to obtain qualitative rather than quantitative data, the eye and the brain are incapable of responding to the demands made upon them. It is in these circumstances that the automation of the microscope really enters into its own. The assessment of the microscopic image has been successfully accomplished recently in America by the work of Ledley and his colleagues. They have developed computer techniques which enable patterns to be recognized and large amounts of significant numerical data to be extracted rapidly from photomicrographs. In this system the photomicrograph, which acts as an intermediary, is analysed by a scanning device which samples the light intensity at 350,000 points in each picture. At each of these points the intensity is expressed numerically as one of seven different levels which are coded and fed into the memory of the computer. If these intensity levels are expressed on a scale where 1 represents white and 7 black, with the intermediate values in between, then a print out of the information in the computer memory will provide a representation of the original micrograph, expressed in digital terms. When the information is fed into the computer the analysis programme begins and the instrument will then produce such information as the number of chromosomes in a spread, the areas of each, the ratios of the arm lengths and numerous other parameters which are of value to the scientist. Such numerical information is available within about 12–20 seconds, whereas if the analysis is performed in the conventional manner by a human operator, then several hours would be required, and the standard of accuracy would almost certainly be much lower.

The introduction of such techniques allows the accumulation of large amounts of quantitative data and will certainly prove to be of great value to the medical scientist who could use such techniques to help him in the screening of large numbers of chromosome spreads and perhaps even of smears which have to be examined for the presence of cancer cells. The logical development of this scheme is to exclude the photomicrograph stage and to link the computer directly to the microscope. This has in fact been achieved in the "Quantimet" image-analysing computer developed in England. This instrument (Fig. 6.13) has a standard microscope fitted with a beam-splitting arrangement which diverts part of the light to form an image on the photocathode of a television camera. From the normal TV scan visual signals can be obtained and displayed on a monitor tube but at the same time, by the use of logic circuitry, it is possible to extract numerical information

FIG. 6.13. The "Quantimet" Image analysing computer. The microscope and the television tube and control unit are on the left; the monitor on the right displays a picture of the field of view being measured.

about the features of the object in the field of view. The output from the TV camera is fed to a detector which responds to changes in the output voltage as the scanning spot passes over the features of varying optical density. The detector provides a pulse of a fixed height but of a length which is dependent upon the time taken for the scanning spot to traverse a particular feature. From these pulses an integrating circuit can measure area and by a very ingenious delay line arrangement it is possible to compare pulses emanating from successive line scans and so pass to a counting meter only one pulse from any object in the field of view. This means that an accurate count can be made of individual features even though they may be traversed by many scanning lines. Various refinements are available on the apparatus; for example the operator can very quickly check visually on the monitor screen the appearance of the field and assess exactly what parameters the machine is recording. It is not possible here to go into details of these new techniques; further references are listed in the bibliography. Up to the present this equipment has largely found its applications in the metallurgical field, for the estimation of the various phases of different metals present in a sample, for the counting of grains and so on, but it seems likely that in the near future the equipment will find application in the fields of biology and medicine. In the early versions of the "Quantimet" the fields to be examined were selected manually by the operator and the various measurements read off in turn from a meter. Future models, now under active development, will provide for entirely automatic sampling by the use of a motor driven "stepping stage" on the microscope and for the automatic recording and print-out of the data which is measured.

Such developments which are now being applied to the microscope have revolutionized our concept of this instrument. The emphasis is changing and now in addition to its routine use for morphological studies, the microscope is becoming a valuable tool for quantitative analysis, providing and with its computer processing, all kinds of information. Although the use of automatic equipment has enabled the operation and handling of the microscope to be simplified so that unskilled operators can obtain large and accurate masses of data, the interpretation of these results still remains a highly skilled operation and presents ever-changing and more complex problems for the scientist.

The comment has often been made in the past that the optical microscope has reached the peak of its development; each time so far the writer has been proved completely wrong. It would be a foolish thing to predict in the present technological climate that the optical microscope has reached the end of the road. There will always be a place for it in the laboratory, both in its present form and in whatever more sophisticated versions are developed in the future.

CHAPTER 7

Greater Resolving Power—the Electron Microscope

ABBE himself realized that the limit of resolution of the optical microscope was set by the wavelength of light, and although he was aware of the possibilities of better results from the use of shorter wavelengths, he was not very optimistic about their successful use. It is ironic that the basic discoveries which paved the way for the eventual development of what must surely be regarded as one of the greatest advances in microscopy since the invention of the compound microscope itself, had already been made by German physicists about 1858 or 1859.

It had been noted by various workers that if electrodes were sealed into a glass tube which was then evacuated, rays which possessed curious properties were emitted from the cathode. Plücker in 1859 showed that these cathode rays were propagated in a linear fashion, whilst ten years later Hittorf proved that they could be deflected by a magnetic field and concentrated by an axially symmetrical field. One other property of these rays which was important for their experimental study, was that although they themselves were invisible, they were capable of affecting a crystalline substance and causing it to emit light. Such substances, often sulphides of metals, are known as phosphors and are much used today in the manufacture of fluorescent light tubes and television screens.

Crookes, working in England, was attracted to the study of these rather mysterious cathode rays and in 1879 he provided a much better proof that they were transmitted in straight lines. It was left to J. J. Thomson, however, working in the Cavendish Laboratory in Cambridge, to provide about 1897 the first theoretical and experimental studies on the quantitative aspects of these rays and to enunciate the concept that they were composed of streams of negatively charged particles, which are now called "electrons".

It fell to de Broglie in 1924 to launch the science of electron optics by introducing one of its most important concepts — that the electron could be regarded not only as a negatively-charged particle but also as a wave. His work was of the greatest importance and it proved a tremendous stimulus to others; it led directly to the studies of Busch who in 1926 showed that the magnetic coil which focused the electron beam could be regarded as analogous to an optical lens. Busch derived the equations for such lenses but he did not put his knowledge to practical use in the construction of a lens.

Although Busch's pioneer work provided the beginning of electron optics and drew the parallel between the trajectories of light waves through a glass lens and electrons through a solenoid, strangely enough it did not suggest that the construction of an electron microscope would be a practicable possibility. This was left to a group of workers in Berlin, including two men whose names are now closely associated with this new development. These were Knoll and Ruska. The Berlin group began a systematic investigation of the possibilities of electron lenses about 1928 and very soon an instrument was built which used an electromagnetic lens to form the image of an aperture in a metal disc although without introducing any magnification.

The next step taken by Knoll and Ruska was to add a second lens and so produce what was in effect the first electron microscope which gave an image magnified about seventeen times. This first true electron microscope (Fig. 7.1a), developed between 1930 and 1931, also used a cold-cathode discharge tube to generate the electron beam which was allowed to enter the tube of the microscope proper through a small hole in the anode. In addition to the concentration coil or condenser lens, a magnetic imaging lens was included, together with provision for an electrostatic lens. This instrument was only capable of forming an image of the cathode of the discharge tube. In the diagram matic section (Fig. 7.1b) the discharge tube, the lenses (which are mounted outside the vacuum) and the transparent fluorescent screen at the bottom of the column may all be seen together with the external photographic apparatus. Development of this apparatus soon enabled emission pictures to be obtained at a magnification of about 400×.

In the early 1930's there was some doubt whether the electron microscope could ever be made into a practicable scientific instrument. It was

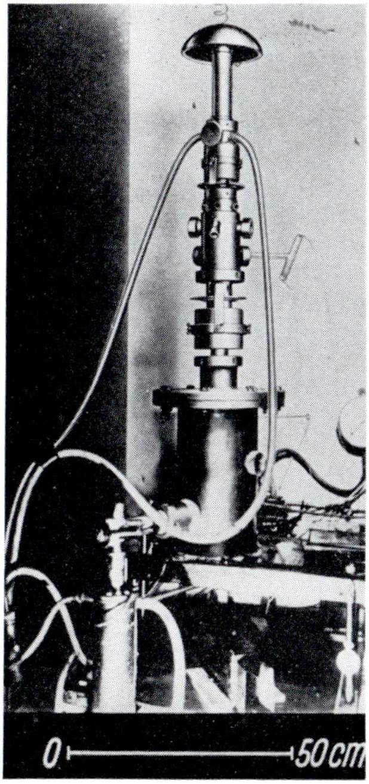

(a)

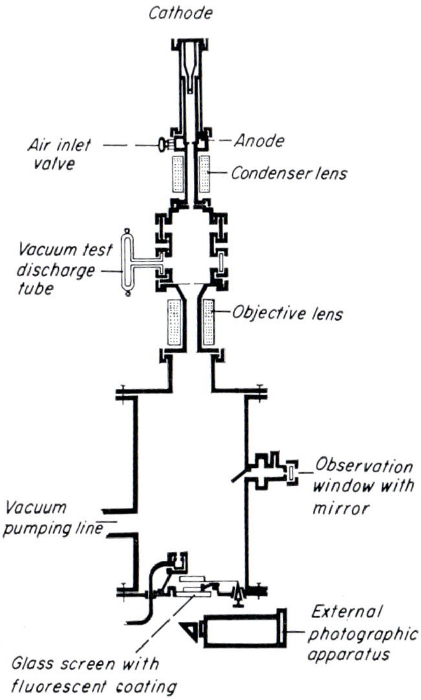

(b)

FIG. 7.1.

(a) Knoll and Ruska's first electron microscope of 1931.
(b) A sectional diagram of this instrument. Note that the photographic apparatus is outside the vacuum system, the image being recorded through the glass fluorescent screen let into the base of the column of the microscope.

true that images had been obtained, but they were only of self-emitting objects; it was generally felt that any attempt to obtain an image by transmission of the electrons through a specimen would result in hopeless failure, either through vacuum damage to the object or to the heating effects of the intense electron bombardment. Difficulties were also foreseen because at that time it was not easy to obtain, let alone maintain, the high degree of vacuum required and to provide for the current in the magnetic lenses to be maintained with the required degree of stability.

Despite these apparent obstacles other workers entered this field and in 1932 L. Marton successfully built and operated a simple electron microscope with magnetic lenses. Ruska also persevered and by 1933 he had built another microscope (Fig. 7.2) which may for all practical purposes be regarded as the direct ancestor of all existing electron

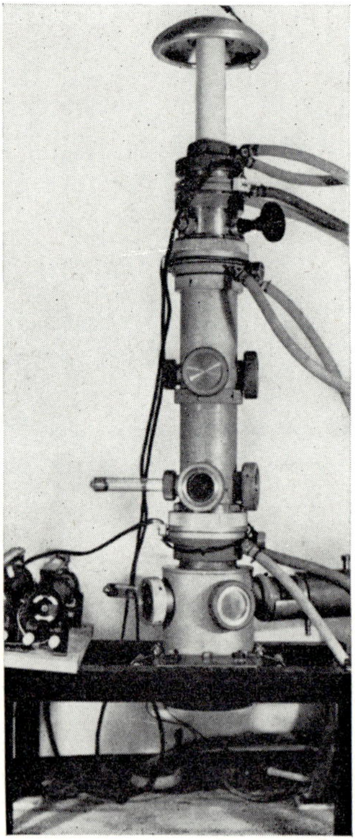

FIG. 7.2. Ruska's electron microscope of 1933. Note the cold-cathode discharge tube at the top and the pipes carrying cooling water to the electro-magnetic lenses.

microscopes. This 1933 microscope retained the cold-cathode tube for generating electrons which were accelerated by 75 kV applied to the anode. There was a single condenser lens and two separate magnifying or imaging lenses, which were known as the objective and projector lenses respectively. This new instrument of Ruska's was designed to accept specimens which were mounted in a horizontal metal disc which could be rotated mechanically (from outside the vacuum) around an axis slightly off-set from that of the column so that each specimen in turn was brought into the electron beam. All the lenses of this microscope were shrouded in soft iron and were provided with water cooling in order to maintain thermal stability; this is a feature still found in present-day electron microscopes. The pipes carrying the cooling water to the lenses are a prominent feature of the microscope shown in Fig. 7.2.

At first the photography of the fluorescent screen was carried out from the outside, but subsequently this was modified and an internal camera was included. With this microscope Ruska obtained transmission micrographs of a piece of aluminium foil and of cotton fibres. Although these were carbonized in the intense beam, the picture still showed a resolution of about 500 Å. In 1934, Driest and Müller working in Berlin with the improved version of this microscope and using internal photography, succeeded in producing photographs of the leg and wings of a housefly at a resolution of around 400 Å. This microscope, although appearing crude by present-day standards, worked and produced pictures by transmitted electrons of objects at a resolution about five times better than it was possible to obtain with the light microscope. At the end of 1934, however, the future of the instrument was still not certain, and research was dropped for lack of immediate support.

At this stage it is convenient to leave the story of the development of this embryonic electron microscope in order to describe briefly the similarities and differences between the electron microscope and the optical instrument and to indicate the special requirements of the electron microscope. Basically the two systems are composed of analogous components; from Fig. 7.3 it can be seen that the light source of the optical microscope corresponds to the source of electrons (the electron "gun") in the electron microscope. Both instruments have a condenser to concentrate the radiation on the specimen, and the primary magnification in both types of microscope is produced by an objective lens.

In Fig. 7.3, the optical microscope is shown arranged to project an image by means of the eyepiece which consists of two separate lenses. In the electron microscope these lenses are represented by the separate intermediate and projector lenses respectively. In the optical microscope it is customary to arrange the system with the eyepiece

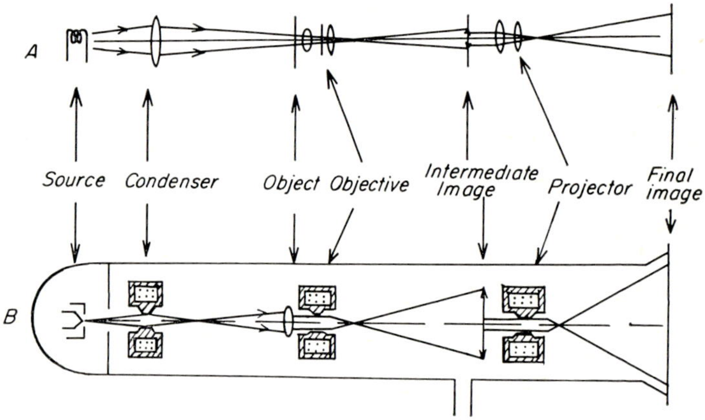

FIG. 7.3. A block diagram comparing the lens arrangements of the optical microscope (a) and the electron microscope (b).

projecting an image upwards onto, say, a photographic plate which is held at the top of the bellows extension of a vertical camera. For reasons of convenience in construction and stability imposed upon it by the long tube, the electron microscope usually has the electron gun at the top of the column and the final viewing screen and the photographic plates are located at the lower end of the column.

Despite these basic similarities which are apparent from the block diagram (Fig. 7.3) there are several fundamental differences. First, as has already been mentioned, the whole column of the electron microscope must be evacuated to a pressure of somewhere around 1×10^{-4} torr (a torr is the unit of vacuum, equivalent to 1 mm Hg) in order that the mean free path of the electrons is increased to about 2·5 metres. This then allows them a clear path down the length of the column (which is usually about 1·5 metres long) without suffering unnecessary scattering on the way by collision with air molecules. Secondly, in the optical microscope the greater part of the total magnification of the system is produced by the objective lens, the eyepiece only contributing in a minor fashion to the final enlargement of the image; this is not so in the electron microscope, where the production of the magnification is more equally distributed between the objective and the intermediate and projector lenses.

In the optical microscope the lenses are, of course, of a fixed focal length and adjustments are made by changing the positions of the lenses, particularly the objective lens, with respect to the specimen. In the electron microscope, however, the situation is exactly the opposite as the lenses are fixed in position and the adjustments of focus and magnification are effected by altering the strength of the lenses by varying the current which is passing through the windings. Again, in the optical microscope the lenses are now highly corrected for chromatic and spherical aberration, as we have seen, but in the electron microscope no such corrections are possible at the present time in the same way as is possible with the optical counterpart.

In present day electron lenses the spherical aberration can only be minimized by accurate workmanship and by "stopping down" the aperture exactly as Hooke did in his microscope of 1665. This, of course, imposes a severe limitation upon the resolution of the instrument. Again, it is only possible to circumvent chromatic aberration by using an electron beam of very high stability, in which all the electrons have the same wavelength, that is, by effectively using "monochromatic" electrons. It is for this reason that the high tension accelerating voltage must be controlled to such rigorous limits. This again, together with the necessity for accurate stabilization of the lens currents in order to obtain a stable image leads to the proliferation of electronic equipment which is such a characteristic feature of the power supply box of an electron microscope, and which contributes largely to the very high cost (up to £20,000) of these instruments today.

A final point of difference which must be mentioned, as it is of importance when one considers the methods for the preparation of specimens for study in the electron microscope, is the actual method of image formation. In the optical microscope the basic image-forming mechanism is one of absorption, in which the object reduces the amplitude of the light passing through it and thereby introduces the necessary contrast. In the electron microscope, however, the contrast in the image is largely due to the differential scattering of electrons by the atoms constituting the actual specimen. The scattered electrons are thrown out of the axis of the instrument and are blocked by coming into contact with the movable objective aperture.

The actual absorption of electrons plays a very limited role in image formation in the electron microscope; the amount of scattering which takes place is very much dependent upon the actual amount of matter present in the object, in other words, by the number of atoms present multiplied by their atomic number. If a heavy metal can be incorporated into the specimen, then the final image contrast will be much enhanced. This may be done by adding the metal either at the fixation stage, by the use of such substances as osmium tetroxide, or at a later stage in the preparation techniques when salts of uranium or lead may be used as "electron stains". As the sections of biological material which are used in the electron microscope are so very thin (being only of the order of 500–800 Å thick), the use of these contrasting agents becomes very desirable, in order to achieve an acceptable image contrast.

The necessity to operate the electron microscope at a high vacuum to some extent proved a limiting factor in the early days of its development, as the production and maintenance of a high degree of vacuum was not then very easy. With the improvements in vacuum technique, in pump design and other technological aspects, this factor has assumed a smaller importance. The other great technological stumbling block in the development of the electron microscope was the need for such accurate control of the accelerating voltage and lens currents. In the early days of the instrument's development (up to about 1940) this often resulted in the proliferation of bulky and inconvenient systems of storage batteries, with all the maintenance problems which the use of storage cells involves. About 1940 the American firm of RCA led the way and introduced electronic stabilization of the power supplies and succeeded in reducing the physical size of the power supply units.

Electronic techniques moved very rapidly during the Second World War and the knowledge gained in this way has since been utilized in improving the power supply and stabilization of the electron microscope.

To return to the further development of the instrument; many workers became interested in this new tool in the late 1930's and it was regarded as an instrument of great potentiality for providing an alternative to the optical microscope for examining objects at a high level of resolution. In particular, L. Marton in Brussels was responsible for the design and construction of a two-stage magnetic instrument. This machine was designed to operate with an accelerating potential of 90 kV and for the first time specimen chamber and photographic plate air-locks were fitted to the column. These, a feature of all present-day instruments, made it possible to change specimens much more quickly and allowed the use of internal photographic plates to become a practical reality.

In this country the initiative was taken in 1935 by Professor L. C. Martin of Imperial College, London, who persuaded Metropolitan-Vickers to manufacture an electron microscope. This instrument, which came to be known as the E.M.1 was described in detail by Martin and his associates Whelpton and Parnum, in a paper published in 1937, although the instrument was in operation during the preceding year. The general pattern of the microscope was modelled on that of Ruska and its design was to some extent conditioned by the desire to examine an object first with the optical microscope and then *immediately* change to examination with the electron beam and study the same part of the same specimen. In order to achieve this the optical microscope was actually embodied in the construction of the electron microscope. Provision was made for the object to be examined with the optical microscope whilst it was in the vacuum of the electron microscope by arranging that the axes of the two microscopes were parallel. The object holder was fitted into a rotating plate which could be accurately rotated by means of a tangent screw; the amount of angular rotation could be checked by observing with a micrometer microscope a graduated circle let into the upper side of the object disc. In this way the object of interest was located in the field of the optical microscope and then by rotating the disc through exactly 180 degrees, the same object came automatically into the axis of the electron microscope. In addition to the tangent screw mechanism which controlled the rotation, there was also a provision for adjusting the position of the object in a radial direction.

The whole instrument stood over six feet high and is shown in Fig. 7.4. The actual column of the microscope is seen in the centre of the illustration, with the cold-cathode discharge tube arising from its upper end, above the expanded part of the casing which contains the rotating object holder plate. The column is strengthened and mechanical vibrations are minimized by the addition of lateral girders, one of which is clearly seen at the side of the column.

The beam of electrons was generated in the cold-cathode discharge tube, which was maintained at a relatively low vacuum, and then entered the actual column of the microscope itself through a hole 0·004 inch in diameter in the aluminium anode. The microscope column was maintained at a high vacuum by means of a large diffusion pump — visible in the illustration at the base of the column in the centre; this large pump was backed in turn by a smaller diffusion pump and a rotary or vane pump. There was a condenser lens and two stages of magnification as in Ruska's instrument; the actual image could be viewed on a fluorescent screen either at the intermediate position, using only one stage of magnification, or by using both stages and observing the picture on the screen at the base of the column. Photography was internal and plates could be inserted into the vacuum of the tube at either of these positions.

Successful photographs were obtained with the E.M.1, operating with an accelerating potential of 20 kV. In the original paper pictures were published of a test object which consisted of a strand of resistance wire 0·0008 inch thick. The photograph taken at the level of the intermediate screen was at a magnification of about 100×, with a resolution of better than 1 micron, whilst that from the final screen had a magnification of 780×. In this case, however, owing to the low power of the projector lens in use, there was considerable pin-cushion distortion evident in the final image.

Although the E.M.1 did not surpass the resolution of the optical microscope, it is noteworthy as the first electron microscope to be made by a commercial firm, and also as an instrument which provided

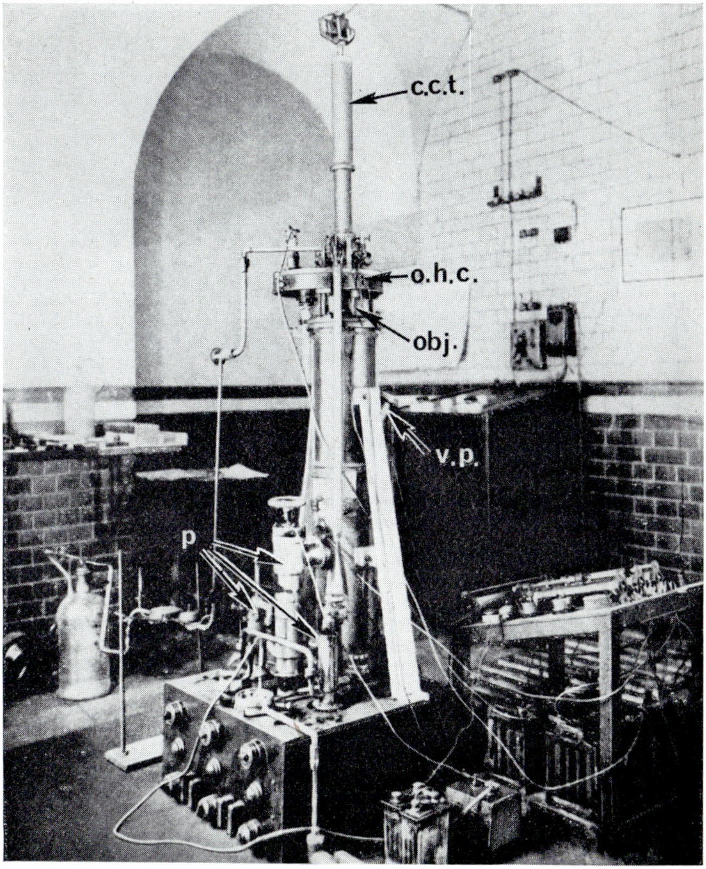

FIG. 7.4. The electron microscope of Martin *et al.* (1936) installed in Imperial College, London. This machine was built by Metropolitan Vickers and was known as the E.M.1. Note the stabilizing girders at the side of the column, the cold-cathode discharge tube (c.c.t.), the object-holder chamber (o.h.c.), the objective lens (obj.), the upper viewing port (v.p.) and the vacuum pumps (p).

valuable experience in this field; unfortunately, owing to the outbreak of war, further developments were postponed for some years.

In Germany, Ruska now working in collaboration with von Borries, had entered into association with the electrical firm of Siemens and Halske to develop an electron microscope for commercial production. They designed a prototype which was in operation by 1938 and the first production model was installed in the following year in the laboratories of I. G. Farben. This microscope is shown in Fig. 7.5; by this time the cold-cathode had been superseded by a heated tungsten filament as a source of electrons, but the stable lens current was still furnished from storage batteries. The general resemblance of the modern Siemens instrument to this microscope of 1939 may be seen by comparing Fig. 7.5 with Fig. 7.8b, which shows the current model, the Siemens Elmiskop IA. It is worth recording that the Siemens electron microscope of 1939 was the first to be placed in serial production and which could regularly surpass the optical microscope in resolving power.

In the early part of 1940 two interesting electron microscopes were constructed. One, described as a "Universal" microscope, was built by Ardenne and was intended to incorporate all the variants in design and function which were current at that time; the other was designed by Marton, now working in America, and was constructed under his direction by RCA. This microscope, which constituted this company's first essay into this new field, was known as the type "A" and never came into commercial production; however, it marked the beginning of the use of highly stabilized power supplies furnished by electronic means, which are now universal features of all electron microscopes. The RCA type "A" was very shortly followed by the type "B" instrument in which the electronic power supplies were made more reliable and were reduced in size. This was so successful that the power supplies could be incorporated along with the vacuum system into the console of the instrument, instead of requiring a separate cabinet of their own. The type "B" microscope, which was largely due to the efforts of Zworykin, Hillier and Vance, proved to be the first commercial instrument available in America, coming onto the market in 1941. It proved to be the forerunner of a long and continuing series of very successful high-quality electron microscopes.

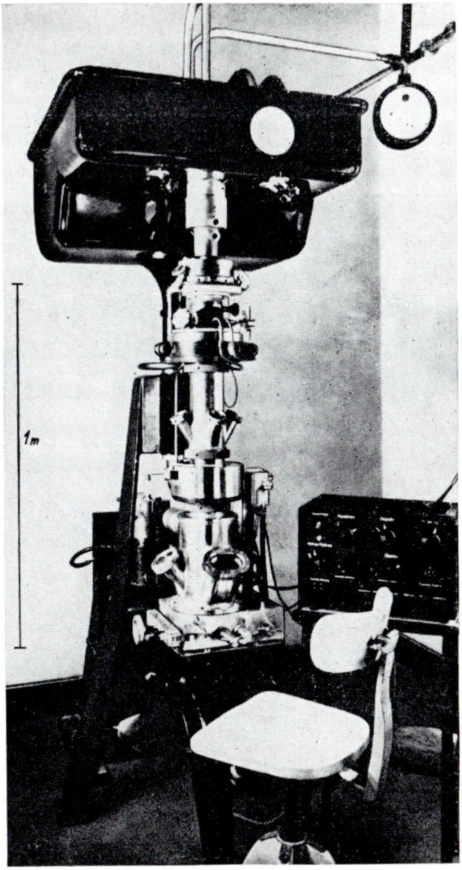

FIG. 7.5. Siemens's first commercial electron microscope (1939).

Progress up to this point in the design of electron microscopes had been rapid. During the ten years or so since Knoll and Ruska demonstrated their first microscope, the main trends had been towards proving the practicability of the instrument and exploring the feasibility of obtaining high resolution by means of electrons. The early electron microscopes were experimental, laboratory-made and often rather crudely constructed; stress was placed on the design of critical operational components and little attention was given to the accessory apparatus or to equipment which did not contribute directly to the production of the image. In consequence, these microscopes were not easy to operate and definitely came into the class of laboratory prototypes which were not suitable for commercial manufacture.

When the future of the instrument seemed assured, and there were indications that a high resolution would ultimately be attained, the engineers began to design electron microscopes in which more emphasis was placed upon the ease of operation, actual reliability in use, and simplicity of construction. The electron microscope was at this stage emerging from a laboratory "string and sealing wax" gadget into the hard world of a rugged, commercially-built instrument intended for long years of reliable service. In this respect the RCA type "B" machine led the way, being particularly advanced in the design of the vacuum system.

At the same time developments in the main-stream of electron microscope design were leading towards the production of instruments with still higher resolutions. One feature which appeared in instruments produced around 1942–3 was developed by Marton and von Borries and Ruska independently. This was the use of three stages of magnification instead of two. In the two-stage microscope the lenses (the objective and the projector) were separated by three to four times the actual length of a lens, with the result that the column was rather long and in consequence there were troubles due to the lack of mechanical rigidity. By the introduction of a third lens (the so-called intermediate lens) the column was rendered much shorter and the versatility of the microscope was substantially increased. With a two-stage instrument the magnification could only be varied over a range of about 5:1 without changing the position of the specimen with respect to the objective, whereas with the newer three-stage microscopes using an intermediate lens this ratio was increased to over 100:1. At the same time the use of three stages allowed the instrument to be set up very easily for obtaining the electron diffraction pattern of a small selected area of the specimen. This feature is of great value in some physical applications as it allows precise information to be obtained as to the nature and configuration of crystalline materials present in the specimen.

In 1944, RCA produced a successor to the type "B" model which became known as the EMU; largely developed by P. C. Smith and R. G. Picard in its early form it had only a two-stage column. The main importance of this microscope lay in the incorporation of many engineering improvements; further refinements were added over the years as they became available. It was claimed by the makers that the EMU microscope gave a resolution of "better than 100 Å", and in actual practice in many laboratories it was resolving about 20 Å.

In England, progress in the development of commercial electron microscopes was delayed by the aftermath of the war, but Metropolitan-Vickers were at work on the design and construction of a new model, the E.M.2, which appeared on the market in 1945. It retained some of the features of the E.M.1, its pre-war ancestor, but was much improved and incorporated the latest advances in vacuum technology such as the use of rubber gaskets rather than the old-fashioned metal-to-metal grease seals. The electrons were generated by a heated tungsten filament instead of from a cold-cathode discharge tube; the high tension and the lens power supplies were also stabilized electronically, a practice which was now universal.

This microscope (Fig. 7.6) incorporated airlocks for the specimen and the camera so that the speed of operation of the instrument was greatly increased. Magnifications of up to 10,000× were available on the screen of the instrument, and further enlargement of the photographic plates was, of course, possible; this microscope was to some extent a compromise instrument in that many desirable features, such as the ability to take stereoscopic micrographs, were sacrificed in the interests of producing a dependable and easy-to-use microscope.

These features were, however, incorporated into the E.M.3 (Fig. 7.7) which was developed about 1947 and came into production around 1949. With the design of the E.M.3 we find for the first time a British microscope designed to incorporate the intermediate lens. In this new instrument the lens windings were now outside the vacuum of the column and mechanical adjustments and alignment controls were provided which worked through vacuum seals. Airlocks were not installed on this microscope, and in order to maintain a rapid rate of specimen and plate change the pumped volume of the column was kept as small as possible. The whole microscope column had to be

FIG. 7.6. The Metropolitan Vickers E.M.2. The electrical controls for the lens circuits are located in the small console to the left of the microscope.

brought up to atmospheric pressure in order to change either the specimen or the plates but the microscope was back at the working pressure within a matter of minutes. It was considered that this method of construction, by avoiding the complication of airlocks, simplified the whole and rendered it more reliable in operation.

The E.M.3 was provided with a continuous coverage of a very wide range of magnifications, with dark-field, with provision for taking

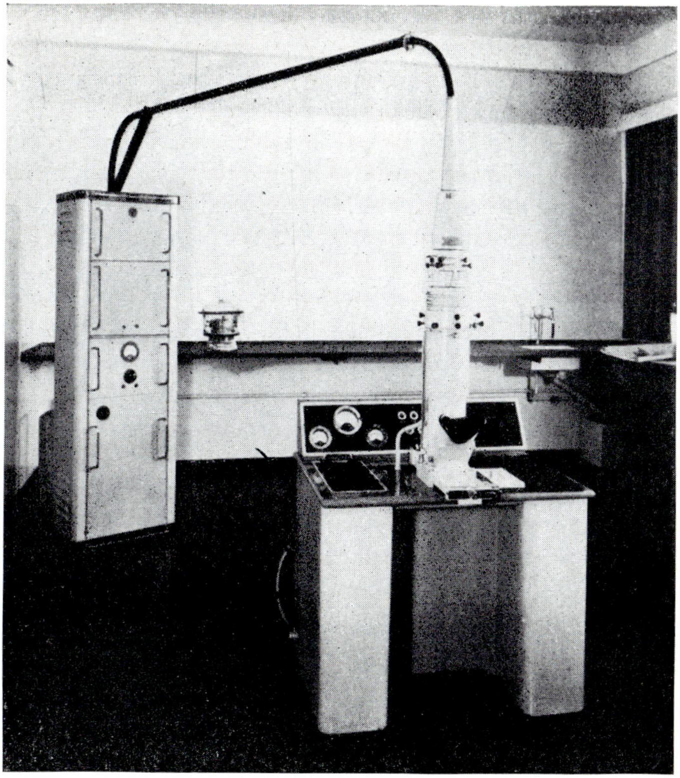

Fig. 7.7. The E.M.3 of 1949. The common convention of placing the microscope column on a desk which also carries the controls has now been established. The cabinet in the background contains the electronic power supplies and stabilizers.

stereo-micrographs and with full facilities for selected-area electron diffraction. The E.M.3 could compete on even terms with the best of the electron microscopes available at that time, and had an average resolution of 35 Å with a best performance of somewhere around 25 Å.

All the early electron microscopes were troubled by the presence in their lenses of the aberration known as astigmatism. It was evident that until such an aberration could be effectively corrected the resolution of the electron microscope was severely limited and really high resolutions would be unattainable.

In 1946, Hillier and Ramberg made one of the significant discoveries in electron microscopy when they discovered the way in which a magnetic lens could be compensated for astigmatism. This was achieved by the so-called "stigmator" which acts by introducing a compensating asymmetry into the magnetic field of the lens in order to neutralize the unwanted errors. Such compensation is now an indispensable feature of every objective and condenser lens and is brought about either by adjusting soft iron correcting slugs on the top of the lens by mechanical means, or in the more refined instruments by applying a potential to one or more of a series of insulated pins or sectors mounted on the lens. The electrostatic field so introduced acts in exactly the same way as a magnetic field and cancels out the unwanted components which are causing the aberration. By this means it is possible not only to compensate for the inherent astigmatism of the lens, but also for that astigmatism which arises during use from the build-up of carbonaceous contamination upon the movable objective aperture. Such deposits are non-conducting and so they charge up under the electron bombardment and they in their turn act as electrostatic lenses and introduce astigmatism into the image.

When the objective stigmator had been perfected as a consequence of the work of Hillier and Ramberg, serious efforts were made to develop the electron microscope to give a really high resolution. In Germany, Siemens had produced in 1950 their model ÜM100 (Fig. 7.8a), which (as its name implies) provided for the use of accelerating voltage up to 100 kV; this instrument was the predecessor of their Elmiskop I (Fig. 7.8b) which was introduced in 1954 and is still current in a modified form. This microscope has a double condenser system, an advantage in that greater intensities can be obtained on the screen at high instrumental magnifications and that a much smaller area of the specimen is irradiated with electrons at any one time so reducing the contamination. It has the usual three stages of objective, intermediate, and projector lens, and was capable of a point-to-point resolution of better than 10 Å, which was at that time a quite astounding figure to be attained routinely from a production microscope.

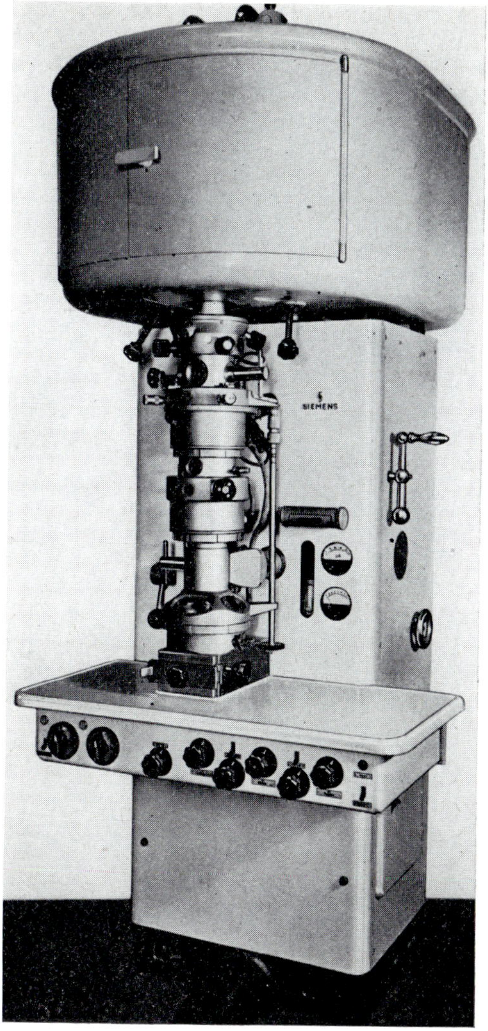

FIG. 7.8.

(a) The Siemens ÜM 100 (1950).

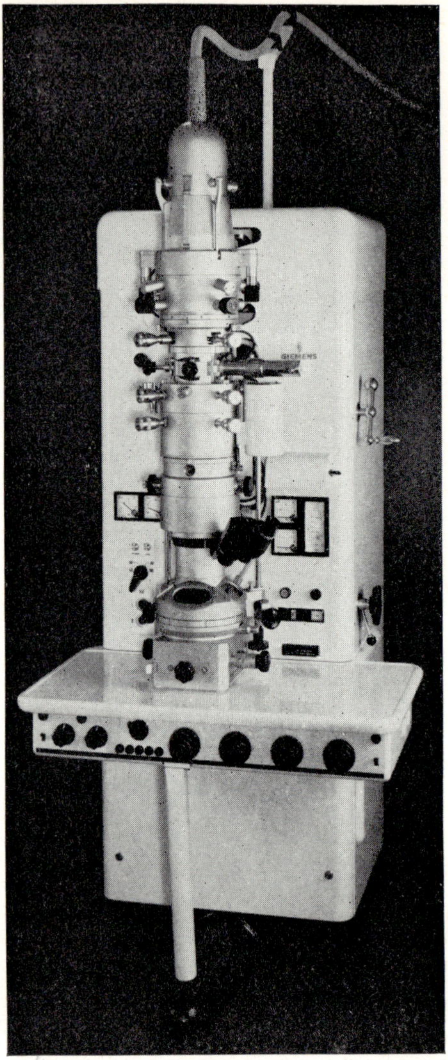

FIG. 7.8.

(b) The Siemens Elmiskop IA (1964). This is their current production model and resembles fairly closely the model I which was introduced in 1954.

(*Siemens Werkbild*)

Two years later the British E.M.6 (Fig. 7.9) was announced; this was a microscope intended to have not only a high resolution, but also to be very versatile, covering all the possible requirements of both the biologist and the physicist, whilst remaining at the same time easy to

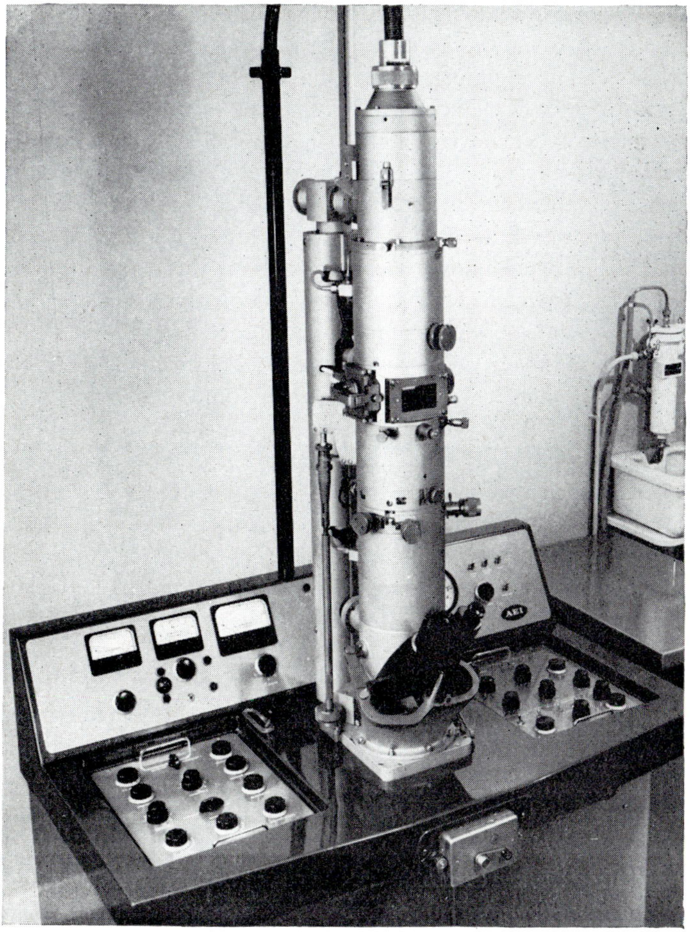

FIG. 7.9. The A.E.I. E.M.6 microscope. The general layout of the column and control panels is well shown in this picture.

operate. Much of the research which had been carried out in the post-war years was embodied in the design of this instrument and it showed a considerable advance over the old E.M.3.*

In particular, the new E.M.6 differed from its forerunner in that it had the double condenser system and an electrostatic beam alignment device which materially helps in the initial adjustment of the microscope after replacing the filament and makes the operation much easier. This same beam alignment device also allows for reflection electron microscopy to be carried out without the necessity of tilting the whole electron gun and condenser system mechanically. Reflection microscopy is especially valuable in metallurgy for the study of the surface of metal specimens and for observing such effects as that of local wear between two metals in contact. The high-tension circuits of the E.M.6 were redesigned and a novel feature at that time was the incorporation of automatic switching so that the lens currents were changed if the accelerating voltage was changed. This automatically refocused the microscope and made it very easy to carry out observations at differing accelerating voltages. The E.M.6 was first marketed in 1958 and although it has now been superseded by the E.M.6G as a general purpose microscope, very many E.M.6 microscopes are giving excellent service at the present time.

Until very recently the general concept in the design of electron microscopes has been towards the provision of a single high resolution instrument which would serve equally well for the biologist, who mainly requires high quality transmission electron micrographs, and for the physicist or engineer. These workers often demand not only transmission micrographs but also reflection pictures, selected-area diffraction, and even facilities for the incorporation of experimental apparatus such as goniometer or heating or cooling stages inside the microscope itself so that the behaviour of their specimens may be studied whilst they are being heated or bent. Designing a universal instrument, as the optical microscope makers had discovered in the eighteenth century, is a very difficult task and only too often something

* It may be noted that the E.M.4 was designed and produced between 1950 and 1953 as a small, compact microscope with a purposely limited resolution, whilst the E.M.5 (which never entered into production) was an experimental high-voltage microscope intended for use with thick biological sections.

has to be sacrificed so that the resulting compromise is not very acceptable to anyone.

This problem has been recognized by Associated Electrical Industries (AEI), the successors to Metropolitan-Vickers in electron microscope manufacture in England, with the result that their most recent instruments are now intended for either the biologist or the worker in the physical sciences. Their general microscope (the E.M.6G) has all the provisions for fitting such attachments as experimental stages and includes a built-in goniometer stage as standard. It has been realized that the operator today is primarily interested in the information to be obtained from the use of the electron microscope and not in carrying out time-consuming and difficult instrumental adjustments. In consequence these modern microscopes tend to have their controls made as simple as possible, and as many as possible of the lens adjustments have been pre-set in order to keep the number of adjustments during actual operation down to a minimum.

The counterpart of the E.M.6G, intended for the biologist, is the E.M.6B; the biologist requires an instrument with maximum resolving power so that structures at the macromolecular level may be examined and also a microscope which will give maximum contrast in the image. This is needed in order to counter one of the chief difficulties of biological electron microscopy which is that the contrast of the material is so often very low. By specializing the instrument it has proved possible to produce a lens of low chromatic aberration (which will in itself tend to improve the image contrast) and to simplify the controls still further, so that the microscope may be used by operators who have had the minimum of training.

Much biological electron microscopy depends on the careful selection of the areas to be photographed, an operation which can only be done by the actual scientist involved in the experimental work. Such scientists often are not really interested in the electron microscope on its own account and so to them a machine which is simple to operate is a great advantage.

The E.M.6B which is now in commercial production is proving a very popular microscope indeed and has consistently resolved image detail of about 5 Å, whilst values of 3 Å can be obtained from suitable specimens.

At the present time there are many high-resolution electron microscopes on the market and it is not possible to enumerate them all here. With any of the current instruments, such as the Siemens Elmiskop IA, the E.M.6B or G, the Dutch Philips EM300, and several of the microscopes now being manufactured by the Japanese, it is easy to obtain instrumental resolution extending down into the range of macromolecular sizes.

The design trends of electron microscopes since 1948 may be conveniently summarized under three headings. First, there was a trend which as mentioned above is now (1966) beginning to reverse. This was the demand for the production of a high resolution instrument which should have the utmost flexibility in operation, or in other words, approach as closely as possible to the "Universal" electron microscope. This demanded a very wide range of magnifications and the provision of a host of auxiliary features, many of which were not used by the eventual purchaser of the microscope. Secondly, there was a trend towards the production of a very simplified electron microscope which was intended for routine use. Such microscopes as the E.M.4, produced by Metropolitan-Vickers in 1954, and the RCA EMC instrument belonged to this category. Although these particular instruments are today obsolete, some, such as the Akashi "Tronscope", the JEM "Superscope" and the small Tesla microscope, might be placed in this category.

The "Superscope" carries the concept of the simplified electron microscope to its ultimate expression. The instrument is of very small size and is entirely self-contained in its cabinet which would almost fit underneath a laboratory bench. It has switched magnifications and a simplified vacuum pump system which has dispensed with the conventional water-cooling of the diffusion pump. The third main trend in the development was perhaps most marked and attracted the greatest amount of attention; this was the production of instruments which would routinely provide the highest possible resolution and image quality. This was initiated by the Siemens Elmiskop I in 1954 and was continued by the E.M.6 and the Philips EM200 and EM300.

It is probably true to say that the very rapid development of the electron microscope from the first instrument of Knoll and Ruska to the accomplishment of modern resolutions of the order of a few Ångstrom units is a triumph of modern instrument technology. It is striking when one considers that a comparable development of the optical microscope took over two hundred years whilst barely thirty years has elapsed since the first instrument of Ruska was operational. It is, of course, this tremendous resolution which constitutes the main advantage of the electron microscope, and justifies the expenditure by laboratories and universities of £20,000 or thereabouts for a modern high-performance instrument. Such resolution had led in biology to a complete rewriting of our concepts of the structure of tissues and cells. This is illustrated in Fig. 7.10a and b. The former represents the appearance of a piece of mammalian pancreas when observed with the highest resolution of the optical microscope. It is evident that relatively little structure can be made out inside the actual cells; there is the nucleus and indications of some granular bodies at one end of the cell but that is about all.

With the electron microscope on the other hand (Fig. 7.10b) the same type of cell shows a wealth of detailed structure in the cytoplasm. Membrane systems fill the bulk of the cytoplasm, and various organelles such as the mitochondria and lysosomes show complex internal organization which was not even suspected with the optical microscope. At very high magnifications (Fig. 7.10b inset) the resolution is such that these membranes of the cell, the "endoplasmic reticulum", are seen to be covered with particles about 150 Å in diameter; these particles are now known to consist of nucleoprotein and are involved in the processes of protein secretion which are carried out by this particular cell type.

Such high resolution is very valuable to the cell biologist but it cannot be attained without some sacrifices, which do to some extent limit the value of the electron microscope in cell biology. First, as the microscope operates at a vacuum of about 1×10^{-4} torr inside the column, it follows that living cells cannot be studied with this instrument. Not only would the water contained within the cell destroy the vacuum, but as it was "boiled off" with explosive force the whole structure of the cell would be disintegrated. Again, as the electron beam has a very poor penetrating power, a thickness of water such as would occur in a living cell would prove to be an impenetrable barrier

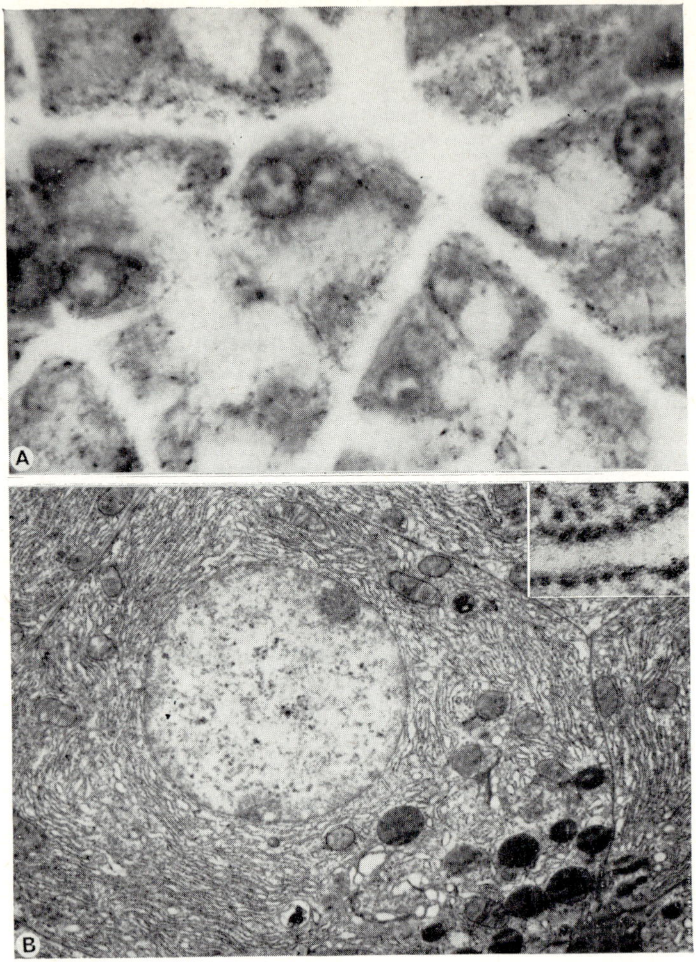

FIG. 7.10.

(a) A high-power optical micrograph of the exocrine cells of the pancreas. Although the nucleus is clearly seen, little structural detail can be made out in the cytoplasm.

(b) An electron micrograph of the same type of cell as in (a) above. Notice that a great deal of detailed structure is now resolved in the cell cytoplasm. The inset shows, at a much higher magnification, the ribosomes which are attached to the cytoplasmic membranes of this type of cell; the resolution of the electron microscope is so great that such magnifications still present us with more useful information.

to the electrons. It is necessary, therefore, to kill the cell, embed it in plastic material and then cut it into very thin slices which are inserted into the microscope as the specimen. The electrons are able to traverse sections of such material of up to 1000 Å thickness and form some sort of an image, but for high resolutions it is necessary to work with sections of material which are considerably thinner than this.

Such electron microscope images suffer from the drawback that they are entirely two-dimensional; if one is studying an 800 Å-thick slice of a cell which may itself be over 40 microns in thickness, it is obvious that in order to survey all the structures present in three dimensions about 480 sections would be required. For various technical reasons sectioning so much material is a very difficult feat, and it is only in very recent years that microtomes have become available which will perform sufficiently well; then each section has to be inserted into the microscope and photographed and the photographs used to make a three-dimensional reconstruction of the object. It is apparent that such a task is full of technical difficulties and has very seldom been attempted.

The lack of penetration of the electrons might be overcome in future years by increasing the accelerating voltage applied to the electrons. It might then become possible to include the living cell in some sort of culture chamber which is outside the vacuum of the microscope column and admit the beam to the specimen via thin windows above and below the chamber. Efforts to develop microscopes working with such high voltages are in progress at the moment, in Cambridge and in Toulouse in France; only time will tell whether it will prove feasible to utilize the tremendous resolution of the electron microscope on the living cell in this particular way.

The electron microscopist, obliged to use only dead material, is always aware of the possibility that the structures which he is examining so carefully are entirely artificial, being produced by the action of the fixative or some other part of the preparative process on the cell. It is very difficult to exclude all such possibilities entirely, as the objects of interest are usually well below the limits of resolution of the optical microscope, so their existence cannot be verified by reference to the direct observation of the living cell. Such indirect evidence as there is suggests that the appearances now current do in fact bear a genuine relationship to the structure present in life, but it is always essential to bear the possibility of artifact formation well in mind.

Again, the living cell is a dynamic entity and what we are examining in the electron microscope is the end product of a whole series of highly artificial operations. These show us the cell as it was *at the time that it was killed* by the chemical fixing agent. It seems not unlikely that as the cell goes through its cycle of activity the ultrastructure would undergo considerable changes. With the restriction on the examination of living cells in the electron microscope the possibility of obtaining accurate information about the dynamic activities of the cell is not very great. It is perhaps legitimate to compare the problem to the difficulty of obtaining reliable information about a cyclical microscopic process by the expedient of taking a series of still photographs at intervals. Such photographs may be strung together in sequence to present some impression of the activity, but the information gained is limited and not comparable with that which could be obtained from a continuous record as might be provided by a cine-camera. With the electron microscope it is only possible to produce large numbers of photographs from material taken at various stages and hope that these can be linked to give some sort of valid impression of the whole process. It is obvious that any development which would allow the examination of living cells in the electron microscope would be an extremely valuable help in the problem of seeking to interpret the relationship between structure and function in the cell.

Finally, in biological work, as has already been mentioned, the electron microscopist is faced with the problem of obtaining sufficient contrast in the image. The use of heavy metals in some form or other does improve the position considerably, but the use of such methods suffers from the drawback that the deposition of heavy metals in the tissues tends to be non-specific; this gives a picture which is, therefore, entirely lacking in chemical information. It is hoped that future developments in specimen preparation technique will lead to the production of micrographs in which the electron densities are directly related to the presence of certain specific chemical groups. By the exact localization of some of the compounds involved in the make-up of the cell a further step forward would have been taken in our laborious process of piecing together the relationship between the chemistry of the cell, its structure and its function.

The very rapid development of the electron microscope itself has been stressed in this chapter; equally rapid and equally important has been the complementary developments and improvements in the methods of preparing material for study with this instrument. The early electron microscopes were much used for the study of particles, either inorganic, such as dust or metallic oxide smokes, or biological particulate matter, such as the bacteria and viruses.

In the case of inorganic particles little more need be done in order to examine them than to arrange for their deposition on a suitable support film which is attached to a specimen grid. In order to add sufficient contrast to biological material, however, it was usual to deposit a layer of heavy metal upon the material by the use of what is called "metal shadowing". The object, mounted on the support film, is placed in a bell jar which contains a tungsten wire filament which carries the metal to be deposited. The air in the jar is pumped out and then a heavy electric current is passed through the tungsten wire. This vaporizes the metal (usually gold palladium alloy or some other heavy metal such as platinum) which settles onto the specimens. If the filament is arranged to lie to one side of the specimen then the deposit of metal is concentrated on one side and this serves to throw the objects and in particular their surface structure into sharp relief. By this method much information about the size and shape of the specimen may be obtained.

At the present time the method has been to some extent superseded by the technique of "negative staining" in which the particle is surrounded by a deposit of electron-dense material dried down from a solution of a substance such as potassium silico-tungstate. The heavy metal contained in the deposit cannot penetrate the particle and so outlines it, sharply delineating any surface irregularities.

The problem of preparing homogeneous material was much more difficult. With thin cells, such as might be found in a tissue culture, some success was obtained by growing them directly onto the support film carried by the electron microscope grids. The cells were fixed by chemicals and then dried before examining them in the electron microscope as whole mounts. By this means Porter was able to show the system of canals (called by him the endoplasmic reticulum) inside the cytoplasm. Later work has confirmed the existence of this system and has emphasized its importance in the transport and synthetic activities of the cell. Such a method of study, however, is limited in its scope and the degree of resolution which can be attained is severely limited by the thickness of the cells. Attempts were, therefore, made to adapt the techniques of the optical microscopist, especially that of embedding the cells in a supporting medium and then cutting them into very thin slices by a microtome.

In 1950, Hillier and Gettner succeeded in obtaining sections 0·1 to 0·2 micron thick by using a modified histological microtome. Such sections, which were obtained from material which had been double-embedded in a mixture of celloidin and paraffin, showed a resolution of about 200 Å when they were examined in the electron microscope. (It may be mentioned, for comparison, that the normal section thicknesses used for optical microscopy range from 5 microns upwards to over 100 microns in special cases.) A great step forward in the art of preparing sections for the electron microscope came when Newman, Borysko and Swerdlow introduced the use of plastics as embedding agents, chiefly using polymerized butyl methacrylate for this purpose. In this way, Palade managed to obtain sections 0·05 micron thick, which showed a correspondingly improved resolution.

About the same time, the first microtome especially designed for cutting thin sections for the electron microscope was designed by Claude and Blum. This machine embodied several of the principles which were found in nearly all subsequent microtomes. Perhaps the most important was the "bypass" principle in which the block containing the tissue to be cut only passed the knife edge on the actual cutting stroke, being deflected on the return movement. Other innovations were the continuous mechanical advance and the use of a trough filled with liquid attached to the knife edge; this received the sections as they were cut and allowed them to form a ribbon which could then be picked up on a specimen grid. From this microtome the famous Porter-Blum model was developed and marketed in 1953. This, together with the introduction some years later of the epoxy resins, led to a tremendous improvement in the quality of sections. These new resins have many advantages over the methacrylates formerly used for embedding materials in order to produce thin sections for electron microscopy. With the aid of the epoxy resins even finer cellular detail could be preserved and pictured, so that the resolutions attained with sectioned biological material improved by leaps and bounds.

It is not possible in a book of this type to enter into all the recent technical advances in specimen preparation for electron microscopy. Methods are changing and improving all the time and progress in this field, although rapid, still lags behind the actual progress in instrument design. One innovation, however, must be singled out for comment and mentioned with the Porter-Blum microtome as contributing to the very rapid spread of biological electron microscopy; this was the discovery in 1950 by Latta and Hartmann that the edge of a piece of broken plate glass could be used to cut much better thin sections than any steel knife. Perhaps more than any other single technological advance this has helped in the preparation of acceptable material for study in the high-resolution microscopes of today.

With the development of the high-resolution electron microscope, the story of the microscope has been brought up to date. There can be no doubt that the end of the road has not yet been reached, even though resolution of a few Ångstrom units, which would have seemed fantastic twenty years ago, let alone at the turn of the century, are now commonplace. There are many fields of active development in optical microscopy, some of which have been indicated in Chapter 6, and even better prospects for the future of electron microscopy. As yet, the design of electron lenses is in its infancy and the search for increased aperture and better corrections still goes on. By this means the resolving power of the electron microscope will be pushed gradually nearer and nearer to the figure of 0·1 Å, which would be expected on theoretical grounds. Alternative methods of recording the image are under examination in which the use of visual focusing on a screen and recording on a photographic plate may be dispensed with in favour of throwing the image onto the tube of an image intensifier and its subsequent presentation by electronic means. Again, quantitation at the electron microscope level is only just beginning.

Something has already been said to indicate the scope for future methods which would yield chemical information at the resolutions of the electron microscope and for methods which would be applicable to the study of the living cell. Variants of the straightforward electron microscope are being developed at the present time. One of the most interesting of these is the scanning electron microscope which is primarily designed for the visualization of the surface structure of an object.

In the conventional transmission electron microscope the electron beam passes right through the specimen; it is the scattering of some of

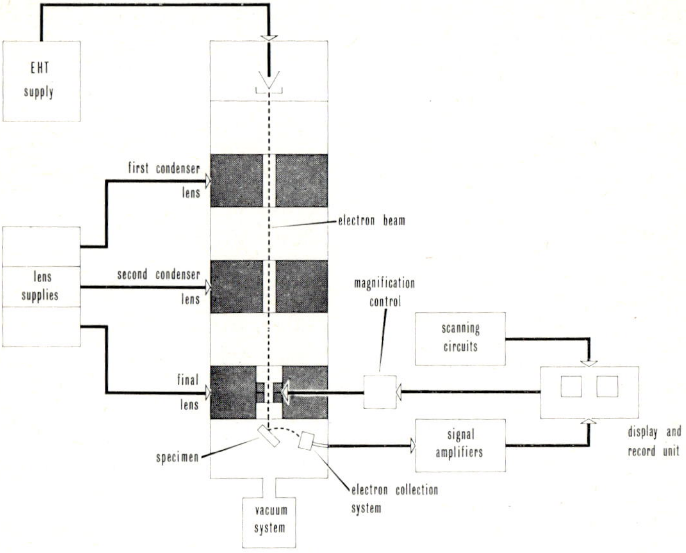

FIG. 7.11. A block circuit diagram of the "Stereoscan" microscope.

these electrons as they pass through the material that gives rise to the image. The scanning electron microscope (shown in diagrammatic form in Fig. 7.11) works on a somewhat different principle; an electron beam produced and focused by a conventional electron gun and lenses, is scanned across the surface of the specimen in a regular fashion by means of a set of scanning coils, rather as the beam in a television tube is scanned in a raster fashion across the face of the tube. When the electron beam hits the specimen, low energy secondary electrons are generated at the point of impact and are emitted from the specimen.

These secondary electrons can be attracted by means of a small positive potential to a scintillator crystal which converts every electron impact onto it into a flash of light. Each of these flashes of light in the scintillator crystal is then amplified by a photomultiplier tube and the final output is presented on a television type of cathode ray tube. This dis-

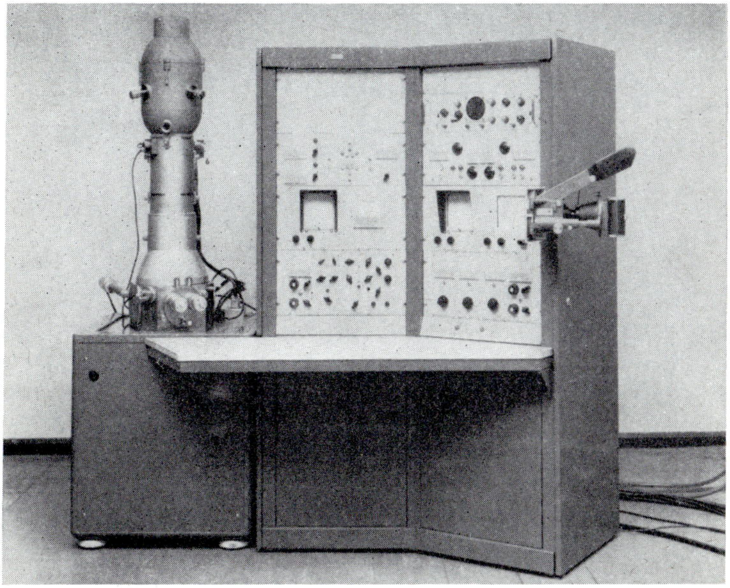

FIG. 7.12. The Cambridge "Stereoscan" microscope. Note the column on the left with the specimen chamber at its base; the controls for moving the specimen can be seen. The console contains the control unit and the display tube, together with the camera unit.

play tube has its beam scan driven in synchrony with that of the original electron-exciting beam so that the resulting image is an exact representation of the surface of the specimen as imaged by the output of secondary electrons. The magnification of the picture can be changed very easily by altering the amplitude of the original scanning beam.

This instrument is made in this country by the Cambridge Instrument Co. Ltd. and is known as the "Stereoscan" (Fig. 7.12); it has not yet achieved the ultimate resolution of the conventional electron microscope but it has advantages which more than compensate for this shortcoming. Perhaps the most important is the fact that the image is that of the surface structure. In the normal instrument it is very difficult to examine the surface of an object without resorting to very tricky technical preparative procedures, with consequent uncertainties in the interpretation of the results. With the "Stereoscan" it is now

FIG. 7.13a. A scanning electron micrograph of the tip of a hypodermic needle. This shows clearly the great depth of field of the scanning electron microscope.

simple to obtain a direct visualization of the surface topography of any object which can be inserted into the specimen chamber and which will survive the vacuum of the electron microscope column. A further advantage of the scanning electron microscope is that it has a very great depth of focus so that all parts of the object appear sharp at the same time. As the image also presents a remarkable three-dimensional character, this is obviously a point of great value when small industrial components are being examined or when a large biological object is under study. Again, specimen preparation for the "Stereoscan" is minimal, being limited to the deposition of a microscopically thin layer of metal on the surfaces to be examined in order to render them electrically conducting.

Many applications of this new tool have already been found; metals, paper, wood, fibres, and ceramic surfaces are all easy to examine whether they are smooth or rough. Biological materials present rather

FIG. 7.13b. A scanning electron micrograph of the inner surface of the dorsal aorta, showing red blood cells still attached to it. Their "spiky" appearance is an artificial change, the true form of a red cell being that of the cell on the upper right of the picture.

more difficulty, as they have to be dried or otherwise prepared so that they will withstand insertion into the vacuum of the instrument. Two examples of the type of picture obtained with the Cambridge "Stereoscan" are shown as Fig. 7.13. Figure 7.13a illustrates the tip of a blunt, contaminated hypodermic needle. This shows in a rather striking fashion what a tremendous depth of focus there is with this instrument. The three-dimensional effect is conveyed rather well by Fig. 7.13b which shows red blood cells attached to the inside wall of a large artery. Many of the cells have distorted during the drying of the preparation of the sample and they show marked irregularities distributed all over their surfaces. Some cells, however, still retain the shape which they have in life. It is certain that the full potential of the scanning electron microscope has only just been realized and the next few years will see many more examples of its use in industrial and biological research.

The scanning electron microscope allows a direct visualization of the surface structure by means of the secondary electron emission; the same exciting electron beam, however, will also cause the emission of characteristic X-radiation from the elements which compose the specimen. Instruments have been designed and built to measure the specific X-rays elicited from each element so that a physical analysis of the specimen may be carried out. Such machines are known as scanning X-ray microanalysers. It is not possible to describe these instruments in detail here, those who wish further information are referred to the book by Cosslett, listed in the bibliography. It is interesting to note again, that, just as with the optical microscope, so we are now moving away from the purely morphological studies with the electron microscope and are adapting these instruments to provide quantitative physical data. Future possibilities in this field are tremendous and progress will be very rapid in the next decade.

The microscope in its various guises has certainly come a long way since it was first seriously used by Hooke 300 years ago for the studies of the surfaces of nettle leaves and of the cut surface of cork. It has long been associated with scientific progress as one of the popular symbols of this craft; in its new roles and with the many rapid and striking developments at the present time it is continuing to live up to this well-earned reputation.

Further Reading

THESE represent a small selection from the numerous works now available on the history and use of the microscope; further references are quoted in the book by Bradbury listed below.

BRADBURY, S., 1967, *The Evolution of the Microscope*, Pergamon, Oxford.

BRADBURY, S., 1967, Some applications of modern microscopy, *J. Roy. Soc. Arts.*

BRADBURY, S., 1967, The quality of the image produced by the compound microscope: 1700–1840. In *Historical Aspects of Microscopy*, ed. Bradbury, S. and Turner G. L'E., pp. 151–173.

CARPENTER, W. B., 1901, *The Microscope and Its Revelations*, 8th edn, ed. Dallinger, W. H., London.

CLAY, R. S. & COURT, T. H., 1932, *The History of the Microscope*, London.

COSSLETT, V. E., 1966, *Modern Microscopy*, Butterworths, London.

HOOKE, R., 1665, *Micrographia*, London.

ROOSEBOOM, M., 1956, *Microscopium*, Leiden.

Name Index

Subject Index